Ainur Yerlan

Functional state of the lymphatic system in circulatory hypoxia

Ainur Yerlan

Functional state of the lymphatic system in circulatory hypoxia

This book is a translation from the original published under ISBN 978-3-659-91569-7.

Publisher:
Sciencia Scripts
is a trademark of
International Book Market Service Ltd., member of OmniScriptum Publishing Group
17 Meldrum Street, Beau Bassin 71504, Mauritius
Printed at: see last page
ISBN: 978-620-3-59741-7

MINISTRY OF EDUCATION AND SCIENCE OF THE REPUBLIC OF
KAZAKHSTAN

RGP INSTITUTE OF HUMAN AND ANIMAL PHYSIOLOGY MON RC

Yerlan A.E. Bulekbayeva L.E.

FUNCTIONAL STATE OF THE LYMPHATIC SYSTEM IN CIRCULATORY HYPOXIA AND ITS CORRECTION

Almaty 2019

Content:

Abstract

Keywords: LYMPHOTICS, LYMPHATIC NUTS, ISHEMIA
BRAIN, HIND LIMB ISCHEMIA, HYPOXIA. CORRECTION

Conclusions. Short-term ischemia-reperfusion of the brain and hind limb in dogs and rats is accompanied by disturbance of rheological properties of blood and organ lymph, which was expressed in shortening of their clotting time, increase in viscosity and thrombogenic processes in blood and lymph. There was an increase in the number of lymphocytes in jugular, lumbar lymph and thoracic duct lymph in rats. Inhibition of spontaneous and evoked contractile activity of cervical and popliteal lymph nodes was found in 70% of experiments. In hind limb ischemia, according to rheography, on the 14th day there was a sharp decrease in tissue blood supply, the amplitude of pulse waves on the rheogram decreased. After 60 days the tendency to normalization was found. The use of Semax, a drug possessing antioxidant, antihypoxic and neutropene properties in cerebral ischemia reduced the negative effect of hypoxia on the functions of the lymphatic system, increased the body's resistance to hypoxia and contributed to restoration of blood supply to tissues. In chronic ischemia of the hind limb (Hour, 7, 14, 30, 60 days) we observed inhibition of lymphodynamics, shifts in morphological and rheological parameters of blood and lymph during the first days and the first month. Spontaneous contractions of the saphenous lymph nodes activated after 30 days, then there was a plateau, but induced contractions of the saphenous nodes in response to the administration of vasoactive substances remained inhibited. Cytoflavin, which has anti-oxygenation and anti-hypoxic properties, had a positive effect on the cellular composition of blood and lymph, pH, clotting time and viscosity of lymph and blood were within normal limits. After 30 days, the morphological composition of the blood recovered, but the hemoglobin level and hematocrit decreased by 15 - 10%, respectively.

INTRODUCTION

Topicality of the problem Disorders in the hemostasis system and rheological properties of blood play an extremely important role in the pathogenesis of vascular diseases, in particular in chronic diseases of lower limb arteries . They account for more than 20% of all types of cardiovascular pathology, which corresponds to 2 - 3% of the total population. The number of patients with lower limb vascular lesions increases with age, making 5 - 7% by the 6th-7th decade of life. (M.I. Lytkin, I.G. Peregudov. 1981.) The disease steadily progresses, worsening the results of conservative and surgical treatment, leading to early disability. At the same time more than 2.7-4.5 million high amputations for lower limb vascular lesions are performed annually in the world.

However, there are few data on the transport function of organ lymphatic vessels and nodes in the hind limb during arterial ischemia of different duration. Meanwhile, these data are contradictory and performed by radiological method using contrast agents. Thus, some authors in acute ischemia of hind limb of dogs described the increase of contrast media transit through lymphatic vessels [Novikov, 1963] and through lymph nodes [Prokofyev, 1976]; other researchers observed a decrease of lymph transport in dogs and cats [Satyukova, 1964; Zorina, 1972]. Histological studies obtained during prolonged hind limb ischemia in dogs lasting from 3 weeks to 2 months showed profound structural transformations of lymph nodes in the zone of ischemic damage [Prokofiev, 1976].

In short-term occlusion of the cranial mesenteric artery in rats, dogs and lower vertebrates a significant decrease in lymph flow and lymph formation processes was observed [Zhumadina, 2006]. Under conditions of venous stasis, when simulating physiological phenomena of weightlessness in terrestrial conditions, the lymphatic system is involved in compensatory reactions of the body aimed at eliminating circulatory disorders and blood stasis [Borodin, Grigoriev, 1986; Bulekbaeva L.E. Demchenko,2004,2007;]. It is known that oxygen deficiency in an organism causes vascular endothelial damage that results in decreased release of relaxing factors and nitric oxide synthesis by endotheliocytes, in an increase of constrictor reactions and vascular permeability and in microcirculatory disorders [Petrishchev, Vlasov, 2000]. In ischemia primary acidosis develops [Jin et al., 1995]. At prolonged ischemia and especially in postischemic period there is an increase of thrombogenic properties of blood, acidosis, increase of free radicals number and development of cellular hyperhydration [Vlasov et al., 2000, Kirheby et al., 2000, Bruce, Austin, 2000]. An increase in lymph flow was noted in dogs when blood oxygen level decreased to 75% [Mayer,1941]. In rats small intestine ischemia was created by mesenteric artery occlusion. The content of free fatty acids in the afferent lymph in mesenteric lymph nodes increased by 20, 6 %, and in the efferent lymph by 1.4

To date, according to the data of our laboratory [Zhumadina, 2003; Bulekbaeva et al. 2005; Zhumadina, Bulekbaeva, 2007], ischemia of individual organs is accompanied by a change in the transs port of lymph along the main vessels: from thoracic and intestinal ducts, which presents a cumulative effect. Due to the important role of the lymphatic system in tissue drainage and its protective and compensatory function in blood flow disorders, it is of theoretical and practical interest to study organ lymph transport through vessels and nodes and processes of lymph formation in a separate organ or region subjected to ischemic damage or interstitial fluid stasis. Conducted researches will allow to study morphofunctional state of organ lymphatic vessels and nodes during ischemia of organs and tissues.

1. Neurohumoral regulation of lymph circulation and homeostasis

One of the important principles of the evolution of the mechanisms of neurohumoral regulation of blood and lymph circulation is the improvement of reactions that ensure the maintenance of homeostasis. It is established that lymphatic vessels and nodes as well as blood vessels are provided with developed sensory and motor innervation [1,2,3]. The mechanisms of neurohumoral regulation of lymphodynamics in higher vertebrates are well enough studied. For example, both an increase in lymph flow [4] and its decrease [5] have been observed during electrical stimulation of the vagus nerve. Narrowing and dilation of the thoracic duct in response to vagus nerve irritation has been described [6]. When the vagus nerve is irritated, the revealed constrictor effect is attributed to the sympathetic fibers going within the vagus nerve [7]. Stimulation of the central segment of the sciatic nerve caused an increase in lymph flow from the thoracic duct, a decrease in the amount of protein in the lymph [8,9], as well as dilation of the thoracic duct and an increase in lymphatic pressure [10].

Changes in lymph flow, biochemical composition of lymph, lymph vessel tone and lymph node volume in mammals when exposed to receptors of internal organs and vessels have been found [1,12,13,14,15,16].

On the basis of observations in patients with neuropathic edemas, connections of these phenomena with functional or organic lesions and various edemas of the brain and spinal cord have been revealed. Anatomical pathways connecting the brain and spinal cord with the lymphatic system have been described, and prelimphatic connections have been established [18,19,20].

In further studies, the role of different parts of the spinal cord and brain in the regulation of lymph circulation and lymph formation in mammals was established [21,22,23,24]. Functional connections of the cerebral liquor spaces with the lymphatic system were revealed [25,26].

It is known that vasomotor innervation has no direct contacts of the end sections of nerve fibers with smooth muscle cells [27,28,29]. The transfer of nerve influences on effector cells is carried out by mediator diffusion; as a consequence, visceral smooth muscle cells do not have postsynaptic membranes, unlike somatic muscles. Therefore, the neural regulation of visceral systems involves a mediator mechanism by the principle of its action on effector structures. There are numerous data in the literature on the influence of acetylcholine, catecholomines and histamine on the activity of cardiovascular and lymphatic systems of mammals. In mammals, catecholamines cause positive and acetylcholine negative inotropic effects on the heart [30,31,32]. The positive inotropic and chronotropic effects of catecholamines on the mammalian heart have been shown to be mediated by different populations of a- and b-adrenoreceptors. A large variability of adrenoreceptor species in the heart has been shown in different representatives of mammals [27,1,29].

It has been established that in blood and lymphatic vessels of different organ affiliation the magnitude of reactions may differ significantly, which is associated with different density of innervation and ratio of adreno- and cholinoreceptors in them, which in turn affect the functional state of the vessel [33,34,35,36,37,38,39,40,].

The data on the influence of mediators and hormones on the mammalian lymphatic system are numerous and contradictory. There have been described an increase in lymph flow from the thoracic duct when adrenaline was injected intravenously [141, 42, 43], its reduction [44], and vessel constriction and dilation [45]. Acetylcholine had a constrictor effect on the cat thoracic duct.

It was found that histamine has a stimulating effect on the heart in rats, guinea pigs, rabbits, cats and dogs [46,47,48,49] and causes a decrease in blood pressure in them [47,50].

Intravenous administration of histamine to adult dogs caused a decrease in arterial and venous pressures, increased lymph flow from the thoracic duct, and increased permeability of the blood-lymph barrier [51]. Perfusion of hyperosmolar NaCl solution to the awake rats resulted in increased arterial pressure. The data support the participation of endogenous histamine in the development of a pressor response to intravenous infusion of hyperosmolar solution [52]. It has been shown that administration of histamine to dogs leads to a 2-fold increase in intestinal lymph flow and a 3-fold increase in intestinal lymph protein concentration [53]. In dogs, histamine increased lymph flow and protein concentration in lymph [54], whereas in cats it did not cause an increase in lymph flow and vascular permeability [44]. In pups, histamine caused an increase in blood pressure and lymph flow [55].

It has been established that cranial vessels and cerebral membranes are abundantly supplied with nerve fibers that participate not only in the regulation of cerebral blood flow but also in functional shifts of the cardiovascular system [56]. The largest part of receptor zones is concentrated in the main brain arteries in the area of cerebral vessel bifurcation (carotid sinus) and in the vessels of the circle of viscera [57]. Special receptor formations located in the walls of cerebral vessels that are specifically sensitive to a number of natural vasoactive substances have been identified. Thus, receptor formations responsive to bradykinin, histamine, and dopamine have been identified [58].

It is now established that the reflex influences from the baro- and chemoreceptors of the vessels also extend to the function of the lymphatic system. It has been shown that when adrenaline comes in contact with the carotid sinus chemoreceptors, the lymph flow decreases [59]. Under the influence of acetylcholine and adrenaline on the carotid zone, changes in lymph flow and respiration were observed in pups from the 2nd day of birth. An increase in lymph flow under the influence of adrenaline was found from the 4th day of the pups' life [60].

When acetylcholine was injected into the internal carotid artery of dogs, there was an increase in arterial pressure and lymph flow [61]. Intra-arterial injection of catecholomines caused lymphatic vasoconstriction in dogs [62].

In addition, biologically active substances, have a certain effect on the motility of the lymphatic vessels. In isolated sections of rat thoracic and intestinal lymphatic vessels, mammary lymphatic vessel and dog jugular lymphatic vessel an increase in spontaneous contractile activity under the action of adrenaline was found, acetylcholine and histamine; high concentrations of these substances caused a tonic effect [13,63,64,65,66,67] The physiological effects of adrenaline and acetylcholine are mediated through the a and c - adrenoreceptors, M - choline receptors, H1 - H2 - histamine receptors, respectively.

Phasic contractile activity of rat lymphatic vessels is carried out with the participation of calcium ions from intracellular sources. Calcium ions transport under the action of vasa-active drugs is carried out by means of the potential of controlled Ca channels [68,69,70].

In ring preparations of bovine mesenteric lymphatic vessels, endothelin ($10\text{-}10 - 10^{-9}$ M) caused a dose-dependent increase in the frequency of spontaneous contractions [71].

Studies of contractile properties of pial arteries in mammals have revealed high sensitivity to noradrenaline, adrenaline, histamine, serotonin and to their fast stretching by potassium ions. Veins, unlike arteries, are not sensitive to mediators and to fast stretching by potassium ions [72].

Isolated human internal carotid and middle cerebral artery strips showed spontaneous electrical and contractile activity of smooth muscle cells. Human cerebral vein strips had no spontaneous activity and also did not respond to mediator influences [73]. Acetylcholine, when acting on isolated human and canine cerebral arteries, causes dose-dependent dilatation, which is blocked by atropine [74].

At occlusion of the right common carotid artery and the left external carotid artery in gerbils with normal neurological indices the blood flow decreased in the cortex and basal ganglia of the right hemisphere, glucose and ATP decreased, in animals with severe neurological indices the blood flow sharply decreased in the right hemisphere and significantly decreased in the medial part of the left hemisphere, glucose and ATP in the ischemic tissue decreased. [75]

In the study of myocardial stress injury in rats there was hypernatremia, hemolymphatic Na+ ratio remained unchanged in blood plasma, and Na+ content in lymph and lymph nodes on the 1st and 3rd day after myocardial stress exposure was significantly reduced. The K+ level on the 1st and 3rd day decreased by 47.6% and 38% respectively and remained reduced for two weeks. [76]

In pathological processes in the abdominal region there is a disturbance of lymphodynamic and physicochemical parameters of lymph, expressed in changes in lymph pressure in the thoracic lymphatic duct and lymph color. In mechanical jaundice, peritonitis and destructive pancreatitis, as the severity and duration of these pathologies increase, the decrease in lymph pressure in the thoracic lymphatic duct leads to a decrease in the volume rate of lymph outflow.[77]

A study of the interaction between leukocytes and the endothelium of pial venules and arterioles during cerebral ischemia showed a significant difference in changes of leukocyte adhesion to the endothelium of arterial and venous microvascular development of hypoxia. [78]

Intestinal ischemia on the 2nd day showed a sharp increase in leukocytes in the blood by 1.2 times higher than in intact animals, on the 7th day there was a sharp increase in leukocytes by 1.5 times, while the maximum

leukocyte content observed on day 1-4 after ischemia was in direct correlation with the severity of ischemic intestinal damage. By day 25 the leukocyte content in the peripheral blood was normalized. [79]

It has been shown that in inflammation of the pelvic organs, there is a decrease in the number of detected lymphatic vessels, their diameter amplitude frequency of contraction

Observations carried out in humans after 3-14 days of limb trauma, the medial collector showed a decrease in lymphatic movement, weakening of the accumulative function of lymph nodes by 30-40%; in the lateral collector the lymphatic movement speed and accumulative function of lymph nodes increased by 20-25%, but on day 21 gradually approached the norm. In the deep collector the maximum increase was observed by day 21. [80] It is known that the lymphatic system provides the most important basis of human health, ensuring the constancy of its two main components - tissue and water homeostasis. [81]

Disorders of cellular and tissue composition lead to metabolic disorders, microorganisms persistence, neoplastic processes. Disorders of the water environment constancy lead to edematous processes and dehydration states, disorders of mineral and water-salt balance. Violations of extracellular matrix state lead to the development of dysplastic processes, to connective tissue diseases, to the disturbances of functional state of the cells and tissues integrated into it [82].

Thus, under the conditions of ischemia-reperfusion of both hind limbs for 3 hours the development of toxicolymph and endotoxemia was revealed, an increased discharge of lymph into the bloodstream was shown. Preservation of drainage-detoxication function and the ability to regulate oxidative homeostasis by regional to the ischemic limb lymph nodes - popliteal and iliac nodes - was shown. [83]

A number of destabilizing factors increase the volume of interstitial fluid in the roots of the lymphatic system, increasing lymph production. General controlled hyperthermia increases the drainage function of the lymphatic system by increasing the minute productivity of the lymphagiae.[84]

The structural changes in the iliac lymph node occurring in the posthyperthermic period indicate that the functional load is redistributed in the "blood-tissue-lymph node" system to reduce it to the adaptation norm. [85]

It is known that in the dynamics of the experiment the regional lymph node acts as an instrument of lymph and blood redistribution between the drainage channels and participates in the regulation of oxidative homeostasis, modifying the free fatty acid distribution profiles. Under the conditions of the atherosclerosis model, the regional lymph node, preserving its homeostatic and transport functions, assumes the function of cholesterol depot, providing regional detoxification. [86]

1.2 Circulatory hypoxia and its effect on visceral functions

Oxygen consumption by tissues   is one of the most important factors, that determined the evolutionary pathways of the cardiovascular system. [Hochachka , 1997].

All energy transformations in the organism are carried out with the participation of oxygen, so the hypoxic state, which changes the ratio between oxygen consumption and energy expenditure, plays an important role in the formation of adaptive mechanisms at the earliest stages of animal development.

Numerous publications are devoted to the influence of cardiac insufficiency on the organism of mammals and humans. Thus, under hypoxia the minute respiratory volume increases [87]. Increased respiratory volume in acute hypoxia is one of the appropriate reactions contributing to increased transport of oxygen to the tissues of the organism.

Hypoxia reveals hyperventilation, which leads to hypocapnia and respiratory alkalosis [88].

It is believed that reflex effects in hypoxia are associated with the stimulation of chemoreceptors of cardio-aortic and sinocarotid reflexogenic zones, as well as chemoreceptors of the brain [89].

After disconnection of aortic chemoreceptors hypoxia there was observed a drop in arterial pressure, which can be explained by a direct oppressive effect of hypoxia on the vasomotor center, a decrease in cardiac minute volume and dilation of peripheral vessels [90].After carotid sinus denervation and aortic chemoreceptor arc destruction, hypoxia continued to cause arterial pressure increase, which suggests that other chemoreceptors than aortic and sinocarotid ones are involved in these reactions [91].

Denervation of the sinocarotid and cardioaortic reflexogenic zones excluded changes in the lymph flow in dogs during carotid artery clamping. The authors believe that the response of the thoracic lymphatic duct to this exposure is a reflex from the sinocarotid and cardioaortic reflexogenic zones [92]. Dopamine is believed to be a modulator between glomus cells and chemosensory nerve endings [93].

In hypoxia there has been described an increase in minute blood volume and heart rate, rise in arterial pressure [95,96]. The increase of blood pressure in hypoxia is believed to be the result of reflex vasoconstriction [94].

Blockade of any of the four main cervical arteries leads to a pronounced reaction of the cerebral tissue. A few seconds after the arterial constriction there is a drop in the level of oxygen tension in the tissue and an increase in the temperature [Berezovsky, 327].

It is known that creatine kinase (CK) is the main enzyme that characterizes ischemic damage of skeletal muscle. This is related to the peculiarities of energy supply of muscle tissue, in which CK and lactate dehydrokinase (LDH) play a key role in conditions of insufficient oxygen supply. In lower limb ischemia in rats CK and LDH dynamics decreased by day 7 and thereafter there was a tendency to increase after 21 days. [98]

In unilateral ischemia in rats, there was a decrease in ATP levels for 30, 60, and 90 minutes without reperfusion. However, this decrease occurred in both ischemic and nonischemic hemispheres, i.e., unilateral ischemia also led to a decrease in ATP production in both hemispheres. It was probably an adapting response to ischemic stress, as was seen [] in the reduction of MDA levels in both hemispheres of rats after 3 h of focal ischemia by occluding the left
middle cerebral artery. Brain ischemia followed by reperfusion triggers a cascade of molecular events, among them lipid peroxidase . For assessing the level of oxidative stress, Onem et al. observed an increase in its level after 15 min of ischemia and the following 15 min of reperfusion. Damage of membrane permeability due to lipid peroxidation can cause decrease of Na + K + ATPase enzymatic activity in the membrane, release of lysosomal proteolytic enzymes and mitochondrial matrix into cytoplasm, start of intracellular proteolysis and cell destruction, Bas O noted that the induction of ischemia for 45 min during occlusion of both carotid arteries, then after 30 min of reperfusion caused the accumulation of oxidation products, including MDA and nitric oxide (NO), as well as the induction of apoptosis in rats . Some scientists have identified .   ,
that a significant amount of .increase in MDA levels in experiments with ischemia followed by different periods of reperfusion, with high-level lipid peroxidation affected in the period of reperfusion

Serteser et al. in a study on rats presented with 60 min of ischemia by carotid artery occlusion , observed changes in lipid peroxidation, with significantly higher MDA values in the ipsilateral cortex compared with the contralateral one.

Systemic hypoxia in rats lasting for 5 min and accompanied by a decrease in pO2 in arterial blood from 87 to 49 mmHg on average caused a decrease in arterial pressure with a simultaneous increase in resistance in caudal ventral artery. After sympathectomy the initial value of this resistance decreased and proved to be stable under the mentioned hypoxic influence. [99].

In lower limb ischemia in rabbits, it was noted that normal pO2 in muscles does not depend on blood flow, but after arterial occlusion pO2 in muscles correlates with linear with blood flow. [100]

It was revealed that in systemic hypoxia in rats the main dilating substance acting on blood vessels is adenosine. However, the dilating effect of adenosine depends on nitric oxide (NO) content: adenosine is probably produced by endothelium and influences it to increase NO synthesis. It has been shown that the deepening of systemic hypoxia is mainly accompanied by activation of a1-adrenoreceptors, which may be very important in regulating blood flow and O2 distribution in the muscle during systemic hypoxia [101].

It is known that in the first minutes of brain ischemia there is a disappearance of impulse activity of neurons. It leads to a sharp disturbance of its energy supply and activation of the lipoperoxidation process. ([16.17] It has been proved in animal and human studies that hypoxia causes an increase in blood supply to the brain, heart, and adrenal glands [102].

Reduced levels of creatine phosphate [103] and ATP [104] have been described in patients with ischemic insulin [105],

In ischemia and cerebral reperfusion, processes at the molecular level cause such disturbances as shifts in Ca^{2+} - homeostasis in nerve cells, free radical generation, mitochondrial dysfunction, protease activation, changes in gene expression, and inflammatory response [106].

Chronic hypoxia attenuated endothelin-1-induced contraction of the pulmonary artery and vein, but intensified saraphotoxin-6s-induced contraction of the vein. Chronic hypoxia enhanced the relaxing effect of acetylcholine and isoprenoline on the pulmonary artery and vein [107].

There have been described reports on the effect of hypoxia on the lymphatic system in mammals. When arteries are clamped in dogs (common carotid, vertebral arteries) the arterial pressure increase, lymph flow increase, thoracic lymphatic duct narrowing are found [80,108]. According to the authors, pressure shifts in the carotid artery area reflexively change the tone of blood and lymphatic vessels.

A decrease in lymph flow from the thoracic duct and acceleration of blood clotting were found in blood loss in dogs [109].The phase character of lymph flow changes was noted: its increase and then decrease from the thoracic duct and reduction of total protein content in blood plasma and its increase in lymph [110].

In ischemia and reperfusion of rat hind pelvic limbs a decrease in lymph osmotic pressure and increase in serum osmolarity was described [111]. During occlusion of the unpaired aortic branch and reperfusion there were noted structural changes in hepatic lymph nodes: they became fragmented from compact ones. The content of free fatty acids in the afferent lymph in mesenteric lymph nodes increased by 20, 6%, and in the efferent lymph by 1.4% [112].

Sclerotic transformation in the cortical and brain substances of inguinal lymph nodes and replacement of lymphoid parenchyma by connective tissue were found in patients with varicose veins of the lower extremities [113] The relationship between separate parts of hemocirculation and lymphocirculation was established, which manifested in multidirectional reactions of resistive, capacitive and lymphatic vessels of different regions during formation of adaptive reactions to hypoxia [114].

Common carotid artery occlusion and blood loss in dogs combined with vagotomy increased arterial pressure and caused lymphatic vasoconstriction. Adrenaline increased lymphatic pressure in a direct dose-dependent manner. These data indicate, according to the authors, the presence of intrinsic regulatory mechanisms in the lymphatic vessels [Daney tn all.,1988] After ischemic cardiac injuries the lymphatic channel plays an important role in the drainage of the necrosis focus [115].

It is known that the autonomic nervous system acts on the smooth muscles of the cerebral vessels by means of mediators, the stocks of which provide prolonged excitation of the nerves [116].

It was found that the human internal carotid artery and the rat portal vein cannot phase contractions without extracellular calcium, although they are highly sensitive to hypoxia and hypercapnia. Lymphatic vessels are less sensitive to extracellular calcium content and are more resistant to hypoxia than venous and arterial vessels [117].

Under hypoxia, complete inhibition of the contractile activity of the thoracic duct smooth muscle cells (SMC) occurs after 1-1.5 hours, of portal veins - after 10-15 minutes, which indicates great resistance of the thoracic duct SMC to a decrease in the partial pressure of oxygen [118].

On rat thoracic aorta ring preparations subjected to chronic hypoxia and on preparations from animals kept under normal conditions it was found that against the background of exposure to No-synthase inhibitor (I-NNA) and heme oxygenase inhibitor (GO) - ZnPPIY the contractile activity of rat aortic preparations subjected to hypoxia to phenylephrine was enhanced compared to that under exposure to I-NNF alone. [119].

Hypoxia was found to induce relaxation of vascular smooth muscle by inhibition of L-type Ca2 channel. In thoracic aortic rings of rats freed from endothelium contracted by phenylephrine, slight to moderate hypoxia caused significant relaxation of the drug. It is concluded that hypoxia acts directly on vascular smooth muscles, causing relaxation partially by inhibition of Ca2 - L-type channels [120].

In rat mesenteric arteries, deep hypoxia caused vasodilation, which was independent of the endothelial state and changes in Ca2 content in smooth muscle. Induced by reoxygenation, vasodilation was caused by a decrease in intracellular Ca2 concentration [180]. Endothelial cells are little sensitive to hypoxia and are less damaged by ischemia than other cells. In postischemic reperfusion, the functional properties of the endothelium, which participates in the development of distant ischemic adaptation, change [121].

A sharp decrease in ATP content was detected in kidneys that were not subjected to reperfusion and the growth of decay products: adenosine, inosine, and hypoxanthine. Cellular loss of adenosine as a result of its degradation during ischemia reduces the pool of adenine nucleotides, which leads to damage of renal tubule cells due to impaired cellular metabolism. On the other hand, accumulation of hypoxanthine during ischemia increases the production of reactive oxygen species (ROS) during reperfusion. [122] Postischemic damage becomes catastrophic when after blood flow is restored the production of ROS increases in an avalanche-like manner, leading to cell necrosis or apoptosis depending on the depth of ischemic stress. *[123]*

A number of studies have shown that the resistance of internal organs to the damaging effects of ischemia/reperfusion can be achieved by ischemic pre- and/or postconditioning. [124]

At the stage of early reperfusion along with blood circulation restoration it is necessary to provide optimization of cellular metabolism, because during this period there are changes in energy supply of cells, disturbance of intracellular ion homeostasis and intensive generation of AOS. Taken together, these factors reduce the ability of cells to restore oxidative phosphorylation, damage the structure of mitochondrial and plasma membranes, causing the so-called "oxygen paradox" effect. [125]

Left ventricular myocardial ischemia caused decreased vena cava blood flow and venous return in cats, which resulted in decreased blood flow and pressure in the pulmonary artery. Atrial pressure shifts had no effect on changes in venous return. [126]

At 10-minute coronary artery ligation, the content of cyclic adeisine monophosphate and cyclic guanosine monophosphate increased both in the ischemic zone and in the nonischemic myocardium of nonadapted rats. After 10 minutes of reperfusion, only the level of cyclic adeisinmonophosphate increased. In myocardium of stress-adapted rats, the level of cyclic adeosine monophosphate increased in response to coronary occlusion, but this increase was significantly lower than in the control group. The authors believe that in adapted individuals, the weak response of the myocardial cyclic nucleotide system in response to ischemia and reperfusion may be directly related to the antirhythmic and cardioprotective effects of adaptation.[127]

After vessel ligation the characteristic clinical and morphological changes developed in the colon. Humoral component in dogs changed towards hypercoagulation and suppression of fibrinolytic activity. Blood coagulation time shortened by 22,3% in 10 m after the blood supply reduction, and by 30,2% in 60 min. Plasma tolerance to heparin increased by 31.2% after 10 min of ischemia and by 35% after 60 min, and antithrombin III content decreased respectively by 31.6% and 24.5%.[128]

2. Blood supply to the lower limbs of animals with ischemia of different duration.

Transcranial Dopplerography (TCD) is a safe and clinically useful technique for assessing cerebral hemodynamics. Observational TCD was first performed in 1982 on intracranial arteries . [129] In general, the advantages of TCD lie in its repeatability, mobility, noninvasiveness, and its timely calculation. TCD can detect intracranial hemodynamic disorders such as arterial stenosis, arterial occlusion, and microembolism. Laser Doppler is sensitive to measure peripheral blood flow, to assess the development of functional collateral vessels. [130]

TCD of lower limb arteries is a valuable noninvasive diagnostic method in the pathology of vascular diseases and an important step in diagnosis and for obtaining subsequent hemodynamic and morphological characteristics. The method consists of image analysis and real-time Doppler information analysis. Doppler information based on the Doppler effect can be used to determine pulse waves i.e. flow velocity shape (hemodynamic characteristics). Spectral analysis is the most important element of the Doppler study of the peripheral arteries of the lower extremity. Based on spectral analysis, the iliac arteries are examined at 3.5 MHz, other lower extremity peripheral arteries are examined at 7, 5, or 5 MHz transducer.[131]

The magnitude of hyperemia detected depended on the duration of arterial occlusion and measurement of time after occluder release. After ligation of the common iliac artery, at the time when blood flow was restored, there was a steady deficit of blood flow capacity reserve, which persisted for at least 14 days. Thus, the use of noninvasive vascular occluder and laser Doppler imaging presents a sensitive and consistent technique for measuring peripheral blood flow for function assessment. These findings will enhance the ability to effectively investigate pharmacological therapies to promote growth and development. [132] In studies of local cerebral blood flow during carotid artery occlusion in rats by laser method on each side, a 0.1 -mm Doppler transducer was performed by moving 0.1 -mm steps across the brain surface. Local cerebral blood flow was expressed in units of LD . The scanned data were used to calculate histogram frequency with flow width classes 5 and LD range from 0 to 150 LD. The observed frequencies were calculated mathematically.

After a stabilization period of 15 minutes, baseline scans were performed on the two hemispheres. The data stream was saved online to the computer. Each scan lasted about 7 minutes. The left and right were occluded . During this prodrome, the head was fixed in a stereotactic frame, thereby ensuring identical positions of the scanned point for subsequent measurements. In the control group, the sutures were not tight and were removed at the end of the experiment.[133]

Observations made on patients with mild traumatic brain injury, with craniocerebral trauma on the background of osteochondrosis-Doppler study of the spine and internal carotid artery revealed symmetrical and asymmetrical arterial constriction borderline with pathology in almost 40% of subjects, and tortuosity pathology in 12%. This proves impaired blood flow in the vertebrobasilar basin.

It is known from the literature that blood flow decreases on day 7 after surgery, reaching the lowest level at 4 weeks, recovering by almost 90% at day 49. Blood pressure, which was lowest on day 14, recovered on day 49 [31] Successful mouse limb revascularization has been shown to occur within 30 to 35 days after induction of severe hind limb ischemia. [32]

Administration of L-arginine at a dose of 30 µ/kg once a day against the background of ischemia led to an increase in microcirculation by 28 days of the experiment, and at a dose of 200 µ/kg once a day led to complete recovery of limb muscle tissue by 28 days.

Laser Doppler perfusion (LDPI) was used to obtain to subsequent reduction of hind limb blood flow, which usually persists for 7 days. LDPI studies showed that hind limb blood flow was gradually restored over 14 days, eventually reaching normal levels occurred between 21 and 28 days. The normal value of the LDPI index was 1.00 ± 0.03 in this study .[34/35]

It was found that proximal femoral artery ligation in the lower limb of the mouse resulted in an 80% reduction of arterial blood flow within 24 hours after surgery compared to the normal contralateral lower limb. Starting from about 48 hours after the active recovery phase, the blood flow in the ischemic limb began to increase. As a result, blood flow stabilized at about 66% of the contralateral non-ischemic limb. Blood flow in the ischemic limb remained relatively stable from 14 to 28 days after injury. Laser Doppler imaging of the ischemic hind limb after surgical femoral artery ligation was performed. Mice from 6 to 8 weeks of age underwent surgical femoral artery ligation. Blood flow was decreased. There were insignificant changes in blood flow restoration within 8 weeks after arterial ligation. These findings have been previously studied .[36 Doppler color was found to be 86% sensitive, 100% specificity, and 97% accurate in diagnosing torsion and ischemia in the painful scrotum. Doppler color is an accurate, non-invasive means of rapid assessment of testicular perfusion in the painful scrotum[37] .Red indicates normal perfusion and blue a marked decrease in ischemic posterior limb blood flow. *[38]*

In rabbit studies in lower limb ischemia it was shown that blood flow pulse after four months of ischemia in the peripheral limb decreased by 15 cm below the level of arteriovenous anostomosis.

When comparing chronic ischemia and acute lower limb ischemia, the recovery of cutaneous blood flow was slower and less complete than in acute ischemia 0.66 ± 0.02, in chronic lower limb ischemia 0.76 ± 0.04, P <.0.5. Chronic ischemia caused blood flow, as measured by laser Doppler scanning and the supply of oxygen in the muscle gradually decreased over 1 to 2 weeks after surgery. [39]

Under normal conditions muscle pO2 is essentially independent of blood flow, immediately after arterial occlusion pO2 correlates linearly with blood flow. Within two weeks of occlusion pO2 is restored to 45% of the baseline level. [40]

In the study of blood flow after complete occlusion, no spectra were observed. For partial occlusion, there was an increase in systolic velocity and a decrease in systolic flow on the distal side of the anastomosis. On the diastolic side, there was a decrease in flow on the proximal side of the anastomosis. [41] On USDG: blood flow in Doppler modes (CFM, PWD, CWD, PD) is not locatable. The lumen of the occluded artery is filled with echomasses of different intensity .[42]

When studying cerebral hemodynamics in patients with a single traumatic nasal hemorrhage, a rheoencephalogram (REG) in the internal carotid artery basin revealed a decrease in the amplitude of pulse waves, a shift to the top of the REG-wave dicrotic tooth and curve rounding, which in general reflected the general trend of hemodynamic changes. In a number of cases arterial tone was unaccountable, as evidenced by the appearance of additional denticles on the catacrota and the formation of "wavy plateau" at the top of the REG curve.

3. Compensatory response of lymph circulation and biochemical composition of lymph in cerebral and hind limb ischemia.

During short-term ischemic brain injury in adult dogs (30 min) and cerebral reperfusion (30 min to 1 hour) caused by occlusion of both common carotid arteries at the level of the neck, lymph flow from the jugular lymphatic vessel on the same side decreased by 25% during the reperfusion period.

The oxygen tension in the blood taken from the external jugular vein was 137 mm Hg in the normal range. During ischemia it decreased and after 60 min from the beginning of cerebral reperfusion it dropped to 78 mm Hg. (Table 1) from the initial level, to 78 mmHg. (baseline -137 mmHg) (Figure 1).

The clotting time of blood and lymph taken 30 min after arterial occlusion decreased by 33 and 30%, and their viscosity increased by 15 and 10%, respectively (Figures 2, 3).

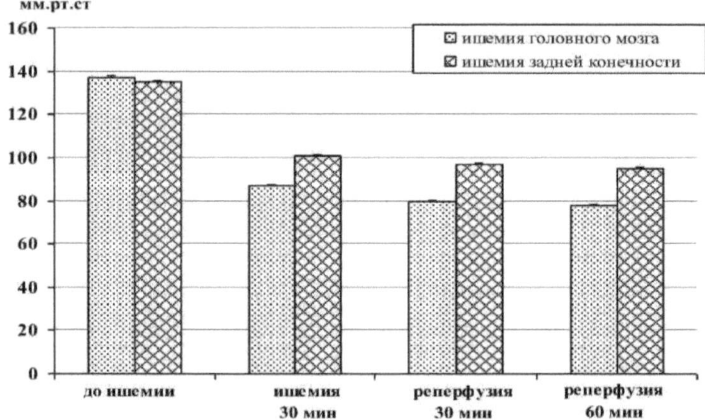

Notes: ordinate axis: oxygen pressure in blood in mm Hg, abcis-experimental stages.

Figure 1 - Blood oxygen tension during cerebral and hind limb ischemia
limb of dogs

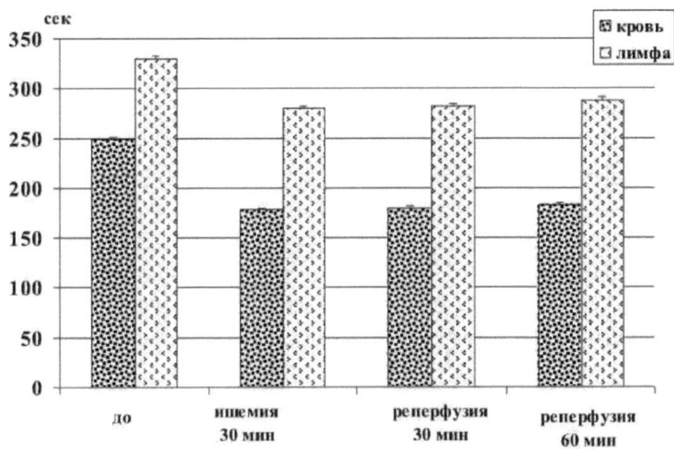

Notations: on the ordinate axis - time in seconds, on the abscissa axis - stages of the experiment.

Figure 2 - Clotting time of blood and lymph during short-term ischemia - реперфузии головного мозга собак

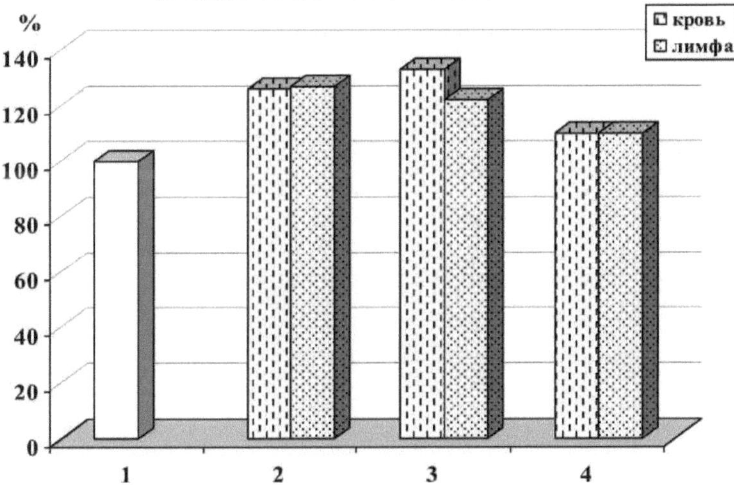

Notes: the ordinate axis is viscosity in percents, 1 is the norm taken as 100%, 2 ischemia within 30 min, 3 after 30 min from the beginning of cerebral reperfusion, 4 after 1 hour of reperfusion period.

Figure 3 - Blood and lymph viscosity during ischemia-reperfusion of the canine brain

During cerebral ischemia in adult dogs, hematocrit (Figure 4), hemoglobin levels in the blood increased from 13.7 to 15.7 g/dl, the number of red blood cells increased from $4.47 \times 10^6 \pm 0.2$ µl to $5.6 \times 10^6 \pm 0.3$ µl (P<0.05) and platelets from $220 \times 10^3 \pm 9.8$ to $300 \times 10^3 \pm 8$ µl (P<0.05).

During the reperfusion period (30 min to 1 hour) after removal of arterial occlusion, blood pH changed toward acidosis from 7.4±0.06 to 7.08±0.05 jugular lymph pH 60 min after the start of reperfusion decreased from baseline to 7.15±0.06 (Figure 5). Lymph generally had a more alkaline reaction than blood (Table 1, Figure 5).

The content of total protein in jugular lymph and plasma decreased after 60 min from the beginning of cerebral reperfusion: in plasma from 67.0±0.13 to 63.41 g/l, in lymph from 27.2±0.32 to 22.0±0.19 g/l. Blood urea content increased, and creatinine content decreased. The concentration of plasma potassium ions increased from 3.8±0.3 to 4.54±0.4 (P<0.05) and calcium from 1.1±0.1 to 1.39±0.3 (P<0.05) mmol/l. In lymph, the level of calcium ions increased from 2.5±0.2 to 4.2±0.5 (P<0.01) mmol/L. The content of sodium ions did not change.

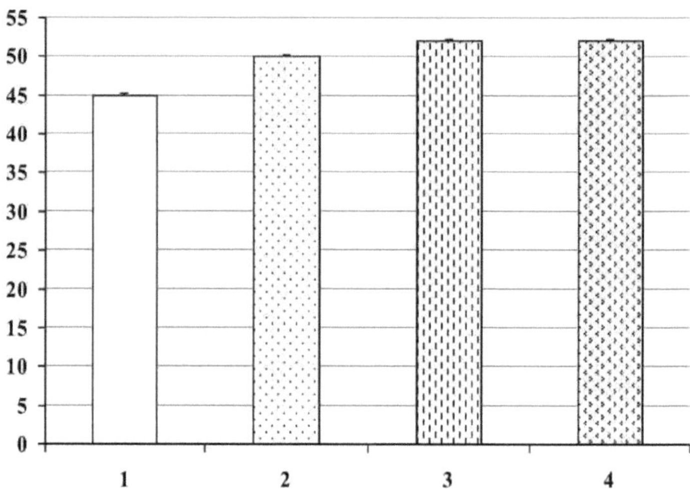

Notes: ordinate axis - hematocrit readings, abcyss axis - stages of experiment: 1 - before ischemia, 2 - ischemia during 30 min, 3 - after 30 min from the beginning of reperfusion, 4 - after 60 min of reperfusion.

Figure 4 - Hematocrit shifts during cerebral ischemia-reperfusion

Rheographic determination of the blood supply to the cervical muscle in rats 30 min after carotid artery occlusion showed a 55% decrease in the blood supply to the tissue compared to the initial background. Vascular pulsation on the rheogram curve decreased.

Changes in the ionic composition of CSF taken from the large cistern of the brain were ambiguous and insignificant The pH of CSF changed toward acidosis.

The leukocytic formula of lymph shows that the cells in the lymph from the jugular lymphatic vessel of dogs mainly consist of leukocytes, and they are 98% to 99%

were represented by lymphocytes and 1 -2% represented by monocytes. The number of lymphocytes in the jugular lymph in the relative resting state was 2320 in 1 mm3 of lymph. After 10 min from the onset of carotid artery occlusion, the number of lymphocytes increased by 28%, and after 30 min, their number increased by 33% from the initial level. And after 1 hour in the reperfusion period of the brain their number began to decrease, and after 30 min their number increased by 33% from the initial level. And after 1 hour in the brain reperfusion period their number began to decrease (Figure 6). The increased number of lymphocytes in the lymph is probably due to their release from the lymph nodes and the need for fluid delivery to the bloodstream due to oxygen deficiency.

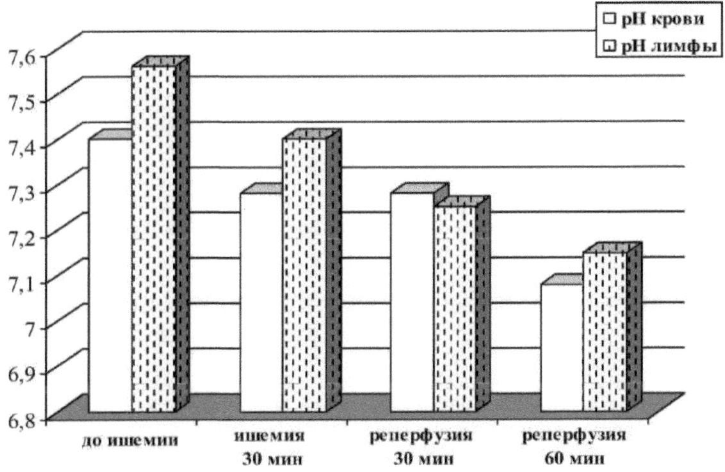

Notations: on the ordinate axis - pH readings, on the abscissa axis - stages of the experiment.

Figure 5 - Shifts in blood and lymph pH during ischemia-reperfusion of the canine brain

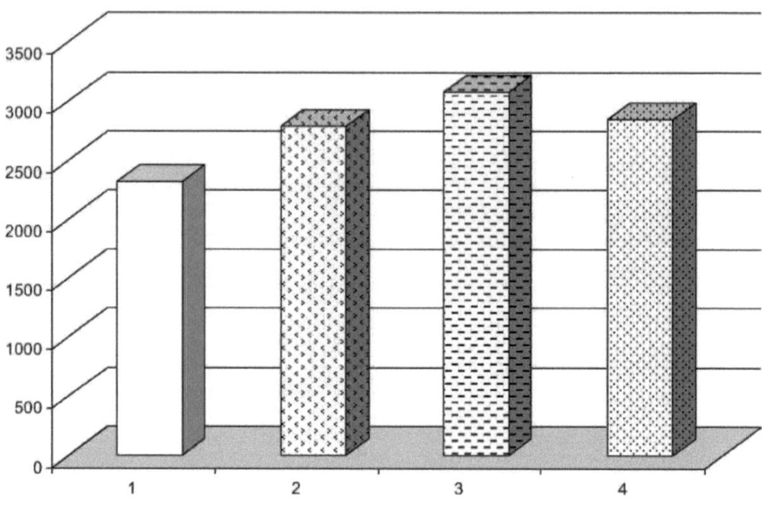

Notes: 1 - normal, 2 - 10 min after arterial occlusion, 3 - 30 min after occlusion, 4 - 1 h after reperfusion.

Figure 6 - Changes in the number of lymphocytes in lymph during ischemia-reperfusion of the brain

The spontaneous contractile reactions of the cervical lymph nodes in rats during cerebral ischemia-reperfusion were inhibited: both the amplitude and the frequency of node contractions decreased. The application of the vasoactive substances, adrenaline, acetylcholine, and histamine at concentrations of 1x10-3 to 1x10-8M to the nodes in question induced contractile reactions of lower magnitude. It decreased by 30% compared to the control group. While the control group showed contractile reactions with increased amplitude and frequency upon adrenaline action, in brain ischemia these reactions were characterized by lower amplitude, however, the curve increased, and their frequency decreased. Under the action of acetylcholine and histamine, both the amplitude and the frequency of cervical lymph node contractions decreased. Consequently, brain ischemia suppresses both spontaneous and evoked contractile reactions of the nodes in the neck region.

Table 1. Physico-chemical parameters of blood and lymph during cerebral ischemia-reperfusion

Title	Before ischemia	Ischemia 30 min	Reperfusion	
			In 30 minutes	In 60 min.
blood pH	7.4± 0,06	7,28±0,08	7.27±0,06	7,08±0,05
lymph pH	7,56±0,08	7,4±0,05	7,25±0,07	7.15±0,06
Blood clotting time, s	248 ± 4,4	178 ± 4,3*	180 ± 3,5*	183 ± 3,6*

	330 ± 5,1	280 ± 5,5*	282 ± 8*	288 ± 7
Lymph clotting time, s				
Blood viscosity, sp	4,2 ± 0,2	5,3 ± 0,3*	5,6 ± 0,4*	5,0 ± 0,4*
Viscosity of lymph, sp	2,2 ± 0,2	2,8 ± 0,3*	2,7 ± 0,2*	2,5 ± 0,2
Hemoglobin, g/dl	13,7± 0,5	14,2±0,7	14,9 ±0,5	15,7±0,8
Hematocrit	45±2	50±4	52±2	52±3

Note - significant compared to control with * P<0.05, ** P<0.01

3.1. Effect of hind limb ischemia-reperfusion on lymph flow and lymph node contractile activity.

The results of the experiments showed that during hind limb ischemia in dogs (30 min-1 h) the lymph flow from the lumbar trunk decreased from 0.025±0.001 to 0.020±0.001 mp/min, by 20% from the initial background. Oxygen deprivation in short-term femoral artery occlusion was weaker than in cerebral ischemia. This was evidenced by a weaker decrease in blood oxygen tension in limb ischemia . (Figure 1) than in cerebral ischemia. During the first 30 min of posterior limb ischemia, its tension decreased from 135.2 mm Hg to 101.7 mm Hg.

The clotting time of blood and lymph decreased by 33% and 25%, and their viscosity increased by 12 and 8%, respectively, as shown in Table 1. During the reperfusion period (30 min to 1 hour), blood pH changed toward acidosis. lymph pH after 60 min decreased from the initial background to 7.22±0.05 (Table 2).

In the period of limb reperfusion, the shifts of these indicators were   more significant and profound, which is clearly reflected in Table 2.

There was an increase in hematocrit, hemoglobin level (from 13.7 to 14.2 g/dl) and the number of blood cells, which we characterize as a compensatory reaction of the blood system in response to oxygen deficiency in the body. The content of leukocytes in the lymph of anaesthetized

Table 2 - Physico-chemical parameters of blood and lymph in ischemia-reperfusion of the hind limb of dogs

Title	Before ischemia	Ischemia 30 min	Reperfusion	
			In 30 minutes	In 60 min.
blood pH	7,4± 0,05	7,32±0,7	7.28±05	7,10±04
lymph pH	7,52±0,6	7,45±0,6	7,35±08	7,22±05
Blood clotting time, s	242±5,4	198±5,3*	188±2,8*	187±3,2*
Lymph clotting time, s	332±5,3	288±5,5*	282±8*	280±7*
Blood viscosity, sp	4,0±0,2	4,5 ± 0,4*	5,0±0,3	5,1±0,3
Viscosity of lymph, sp	2,0±0,2	2,3 ± 0,3*	2,5±0,2	2,5±0,2
Hemoglobin, g/dl	13,7±0,5	14,2±0,7	14,3±05	14,6±08
Hematocrit	45±2	48±3	50±4	50±5

Note - significant compared to control with * P<0.05, ** P<0.01.

was 2330 in 1 mm3 of lymph in the relative resting state and increased after 30 min from the beginning of femoral artery occlusion by 10%, after 1 hour by 20%, and after 3 hours was 2120 in 1 mm3 of lymph compared with and a similar value.

During ischemia-reperfusion of the hind limb of dogs, changes in the ionic composition of blood plasma and organ lymph were observed. During reoxygenation of the hind limb, 60 min after the start of reperfusion, there was an increase in the concentration of potassium and calcium ions in the blood plasma taken from the femoral vein and in the lymph taken from the lumbar lymphatic trunk of the dog. The content of sodium ions in blood plasma and lymph slightly increased.

According to the rheographic index and systolic wave amplitude, hind limb ischemia in rats after 30 min was accompanied by a 55% decrease in blood flow. The pulse wave and magnitude of pulse vascular oscillations decreased.

Spontaneous and evoked contractile activity of the popliteal lymph nodes in rats with hind limb ischemia-reperfusion decreased compared to the control group. The amplitude and frequency of spontaneous node contractions decreased. The expression of contractile reactions decreased and their latency period increased in response to vasoactive agents (adrenaline, acetylcholine, histamine), as shown in Figures 7, 8.

In intact rats, spontaneous contractions of isolated popliteal lymph nodes were observed with a frequency of 3.5±0.4 socrine/min and an amplitude of 7.2±0.3mg. In intact rats, the popliteal lymph nodes contracted with an increase in amplitude by 48% and frequency by 23% from baseline on the action of 1×10-8-1×10-3M adrenaline, whereas on acetylcholine (1×10-8-1×10-3M) the nodes contracted with a 36% increase in amplitude and a 20% (p<0.01) increase in frequency from baseline, respectively. The stimulus threshold for vasoactive agents was 10-8M

In hind limb ischemia, the contractile activity of the patellar lymph nodes decreased by 35-40% on the action of adrenaline compared to the control group. Against the background of hind limb ischemia under the action of acetylcholine, the amplitude of lymph node contractions decreased by 61% and the frequency by 52% (p<0.01). Similar reactions were observed when histamine (1x10-8-1x10-3M) acted on hind limb ischemia. The stimulus threshold for vasoactive in vesicles increased to10-7M.

Thus, in short-term ischemia-reperfusion of the hind limb, there is a depression of the transpiratory function of the lymphatic system, in particular, a decrease in lumbar lymph flow and a decrease in spontaneous and evoked contractile activity of patellar lymph nodes with their decreased sensitivity to vasoactive substances and increased latent period of reactions.

Consequently, oxygen starvation of the hind limb tissues during limb ischemia has a depressing effect on the functional state of the lymphatic system.

In 6 rats with chronic occlusion of one femoral artery under aseptic conditions, the popliteal lymph node was taken after 7 days and histochemical fluorescent microscopic examination revealed changes in the density and area of adrenergic innervation in the popliteal node capsule. Reduction and disappearance of some part of terminal branches in plexuses around blood vessels of the capsule, paler luminescence of nerve fibers were detected.

It can be assumed that oxygen starvation of hind limb tissues, including lymph node cells, negatively affects the structural and functional state of this organ.

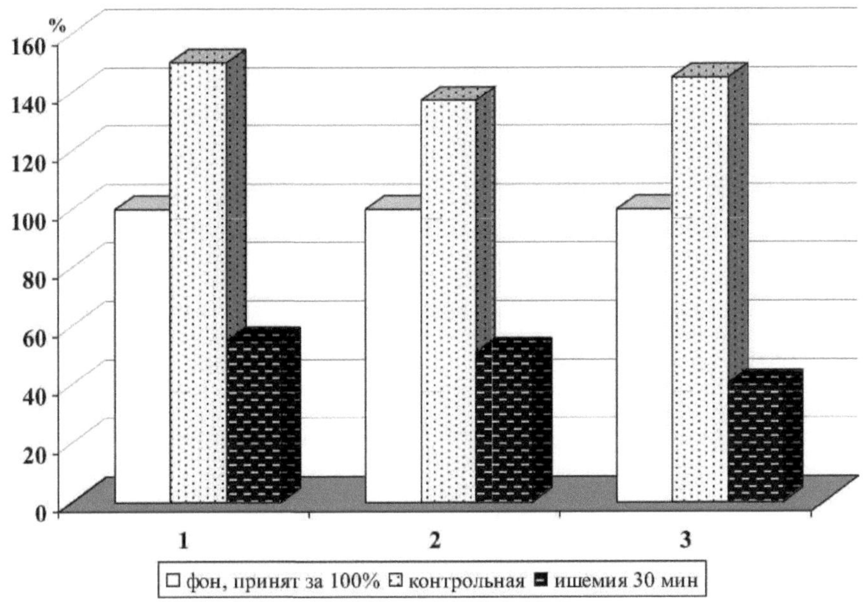

Designations: 1 - adrenaline ($1 \times 10^{-8} - 1 \times 10^{-3}$ Ml), 2 - acetylcholine ($1 \times 10^{-8} - 1 \times 10^{-3}$ Ml), 3 - histamine ($1 \times 10^{-8} - 1 \times 10^{-3}$ Ml).

Figure 7 - Changes in the amplitude of contraction of the popliteal lymph node in rats in norm and in hind limb ischemia under the action of vasoactive substances.

Designations: 1 - adrenaline (1x10-8-1x10-3 Ml), 2 - acetylcholine (1x10-8-1x10-3 Ml), 3 - histamine (1x10-8-1x10-3 Ml).

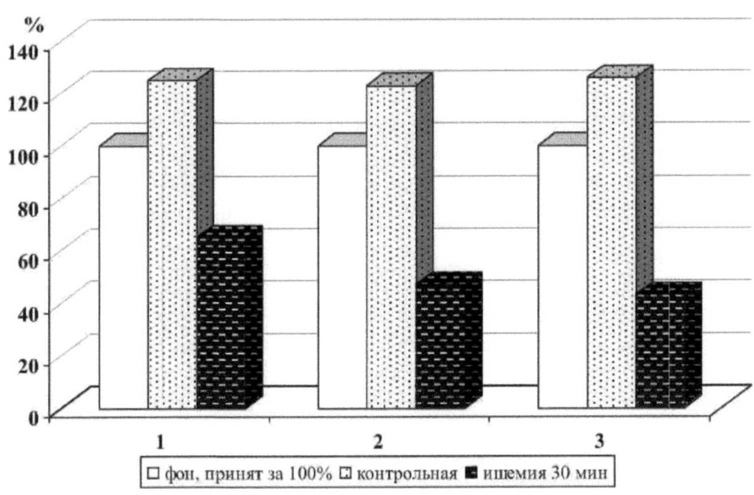

Figure 8 - Changes in the frequency of cervical lymph node contractions in rats in norm and in hind limb ischemia under the action of vasoactive substances.

3.2.Morphofunctional state of lymphatic vessels and nodes in chronic hind limb ischemia

3.2.1 Adrenergic innervation of vessels and lymph nodes, lymph flow and blood flow, biochemical parameters of lymph and blood plasma in chronic ischemia of the hind limb

In rats, 14 days after occlusion of the femoral artery of the hind limb, there was a decrease in the oxygen tension in the blood to 90%. After 30-60 days it decreased to 88%, after 90 days it increased to 98%.

In the tissue of inguinal and subchain lymph nodes, adrenergic innervation predominates, addressed to the blood vascular network. The follicular innervation is formed mainly by single adrenergic fibers. Adrenergic nerve fibers referring to the blood vessels have more varicose thickenings and their fluorescence is higher, which indicates a greater concentration of noradrenaline in these nerve fibers.

Thus, the lymph node tissue receives innervation to a greater extent through the adrenergic innervation of the blood vessels of the lymph node. This type of adrenergic innervation of the tissue is considered indirect. Consequently, the inguinal and popliteal lymph node tissue is dominated by the indirect type of vasomotor innervation (Fig. 10).

The adrenergic nervous apparatus of the femoral arteries and veins of intact rats was represented by a developed innervation consisting of brightly fluorescent fibers with numerous regularly arranged varicosities. Nerve plexuses were arranged in a continuous dense network in the blood vessel wall and fluoresced uniformly along its entire length. The density of adrenergic fibers in the femoral artery wall was higher compared to the vein of the same name. The femoral vein had a large branched nerve network, whereas in the femoral artery wall, the nerve fibers formed a dense small branched network.

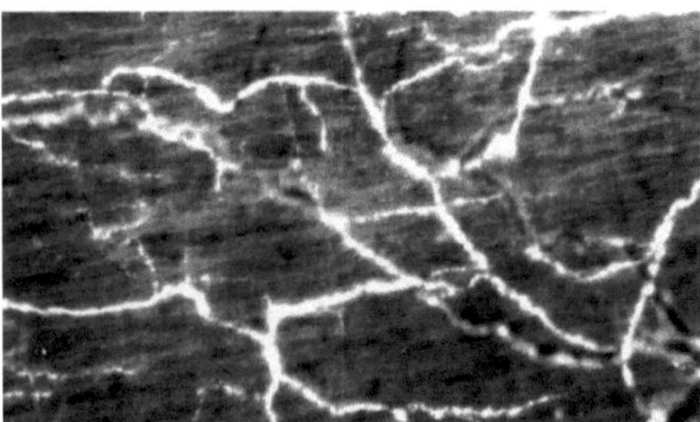

Figure 10- Adrenergic innervation of the popliteal lymph node capsule of the intact rat
O6.30.OK. Rk 6.3.

In rats after 2-week hind limb ischemia, there is a sharp blood filling of the blood vessel network feeding the lymph node tissue in the tissue of the hamstring lymph node. The adrenergic network formed along the course of the blood vessels is preserved. Both nerve fibers and their varicose thickenings, both along the nerve fiber length and at its ends, i.e. terminal varicose extensions have bright fluorescence, indicating a high concentration of norepinephrine in them.

The preparations of the femoral artery of the ischemic rat limb showed the blood filling of microvessels of the arterial wall. Diffusion of norepinephrine from varicose dilations of the terminal part of the nerve fiber was observed.

In the femoral vein wall of the ischemic limb, the adrenergic innervation was preserved, but the luminescence of nerve fibers around the vasorum uasis was more intense.

In rats after 1 and 3 months of hind limb ischemia the follicular single nerve fibers in the tissue of the hamstring lymph node are single, there is no clear near follicular ring. The near vascular adrenergic nerve network is preserved to the extent that the blood vessels in the lymph node tissue are preserved. Excessive enlargement of connective tissue in the lymph node and their nonspecific luminescence are very clearly detected. The preserved adrenergic network in the lymph node tissue has an irregular glow (Fig. 11).

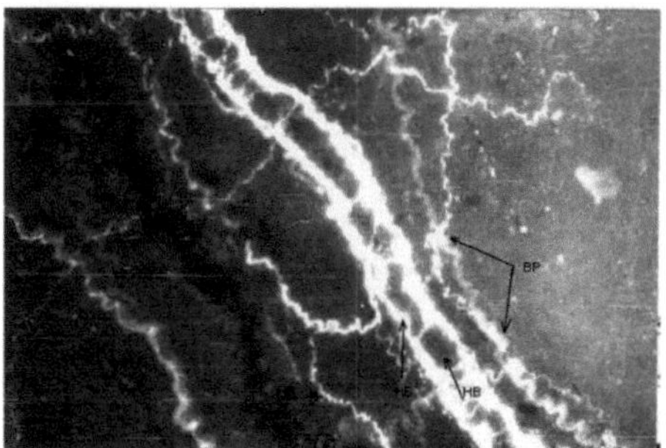

Figure 11 - Adrenergic nerve fibers in the capilla of the popliteal lymph node of rats after 14 days of ischemia

Ob.30.OK.Rk 6.3

We observed the same changes in the structure of the adrenergic innervation apparatus as in the lymph node tissue in the femoral lymph vessel wall in rats with 2-week, one-month, and three-month hind limb ischemia.

Experimental results showed that during hind limb ischemia in rats, lymph flow from the thoracic duct taken before the diaphragm decreased from 14.7±3.5 µl/min/100 g.m.t. control group, 2 weeks after femoral artery occlusion to 8.7±2.9 µl/min/100 g.m.t, after 30 days to 8.3±2.2 µL/min/100g m.t., after 90 days to 7.9±1.3 µL/min/100g m.t.

The content of total protein in the chest duct lymph was 39.2±0.7 g/l, after 2 weeks 28±0.5 g/l, after 30 days 27±0.6 g/l, and after 90 days 27.2±0.3 g/l. Blood clotting was accelerated. In control - 250±14 s, after 2 weeks -176±6 s, after 30-90 days - 191±17 s. Lymph clotting time was also accelerated: control 347±20 s, after 2 weeks 288±19 s and after 30-60 days 308±22 s. Hematocrit increased from 46±3, after 2 weeks to 48±4 and after 30-90 days to 49±3.

The results of biochemical studies showed that chronic hind limb ischemia in rats leads to a decrease in the content of total protein in lymph and blood plasma. After chronic hind limb ischemia urea content in lymph increased by 17%, creatinine - by 15%, AST level - by 25%, ALT - by 50%. In blood plasma the shifts of these indexes were more significant.

In the control group of rats, spontaneous contractile activity of isolated popliteal lymph nodes was expressed as rhythmic contractions with the parameters: frequency - 3.2±0.2 contraction/min, amplitude - 5.1±0.4 mg. The study of spontaneous contractile activity of popliteal lymph nodes in hind limb ischemia revealed a decrease in frequency and amplitude of contractions in 62-65% of experiments. Frequency of contractions of the popliteal lymph nodes was 1.3±0.1 contraction minutes, the amplitude value decreased significantly, almost 1.5 times from the initial values and was - 3.5±0.1 mg. The contractile activity of the popliteal lymph nodes under the action of the vasoactive agents adrenaline, acetylcholine and histamine (1 *10-6M- 1x10-3M) was reduced by 40-50% relative to control. Our studies have shown that after hind limb ischemia in rats there was a decrease in blood plasma and lymph volume, as well as a decrease in the spontaneous contractile activity of the hamstring lymph nodes.

The morphological blood pattern 14 days after femoral artery occlusion of the hind limb was characterized by an increase in all parameters: number of erythrocytes, leukocytes, and platelets by

10-16.5%. Hematocrit increased from 48% in control to 49%, hemoglobin increased by 5%. After 30 days from the beginning of ischemia the increase of these indices continued. After 60-90 days there was a tendency to their normalization, however, hematoglobin content after 90 days decreased from the norm by 25%. Hematocrit decreased to 36 -35%. The number of platelets in the blood recovered.

The lymph flow from the lumbar trunk of the dog decreased by 20%. In 14 days after femoral artery occlusion in the groin of rats, the content of biochemical parameters changed towards their increase. Urea, creatinine, ALT and AST in the blood increased. Especially the level of ALT increased by 150% and AST by 120%. The content of total protein in blood plasma and lymph decreased. In lymph these components were lower than in blood, but the trend during ischemia was similar.

In rats, limb ischemia was performed up to 3 months. Their lymph was taken from the thoracic duct under the diaphragm. Normalization of biochemical parameters was observed only after 3 months. However, the blood picture was still altered. Hemoglobin level and hemotocrit were decreased.

Thus, even after 90 days, oxygen deprivation in the tissues persisted. The number of erythrocytes was slightly higher than normal, but the hemoglobin content per erythrocyte decreased sharply compared to the control.

According to the results of contour analysis of the rheographic curve, which was reduced to its qualitative and quantitative assessment at 2-week ischemia the rheographic index decreased by 20%, at three-month ischemia it decreased by 50% (Table 5).

Table 5 - Rheogram indices in circulatory hypoxia

Rheogram Indicators	intact rats	2 weeks of ischemia	4 weeks of ischemia	12-13 weeks of ischemia
Amplitude of the systolic wave [Ohm]	0,015±0,03	0,013±0,004	0,010±0,04	0,09±0,03
Fast Blood Flow Rate [ohms/s]	0,20±0,05	0,17±0,03	0,15±0,4	0,10±0,3
Slow blood flow rate [ohms/s]	0,15±0,03	0,18±0,01	0,20±0,01	0,23±0,02

The data in the table show that circulatory hypoxia of the hind limb of different duration causes a sharp weakening of blood flow, deterioration of blood circulation in the microcirculatory vessels, reduction of oxygen delivery to tissues, causes tissue hypoxia and tissue metabolic disorders. As the circulatory hypoxia develops, the visual assessment of the limb rheogram graph decreases the amplitude of RVG waves, the rising part of the wave becomes flat, the vertex flattens, additional waves disappear. At 3-month ischemia pulse oscillations changed insignificantly and approached a straight line.

During experimental ischemia we administered cytoflavin after 14 and 30 days, we observed its positive effect on adrenergic innervation. Fluorescence was decreased in the wall of lymphatic vessels in the terminal varicosities of nerve fibers, and norepinephrine diffusion was observed in the arterial wall 1, but with less intensity. Somewhat greater contractile activity of the hamstring lymph nodes was observed in rats with 4 weeks of ischemia . Blood oxygen tension decreased insignificantly after 30 days, by 20-25% of the norm. When cytoflavin was used, the oxygen tension in the blood decreased only by 10 - 15% of the norm.

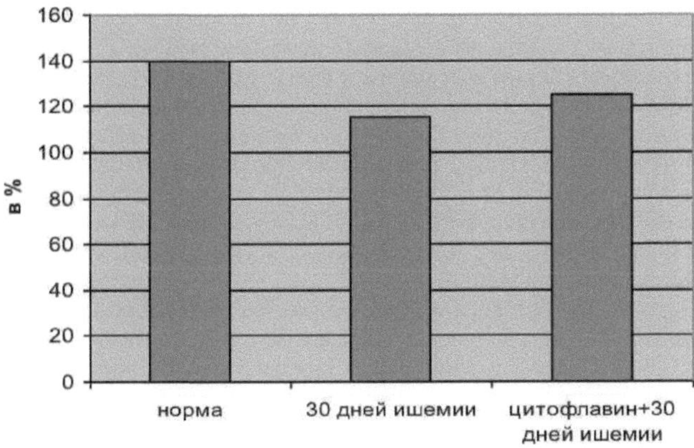

Notes: Ordinate axis - percentage oxygen content in blood, abcis - series of experiments

Figure. 12 Oxygen tension in the blood during prolonged limb ischemia.

3.3 Correction of ischemic cerebral injuries and chronic posterior limb edema
3.3.1 Adrenergic innervation of lymphatic vessels and nodes and in chronic limb ischemia and correction with cytoflavine.

Cytoflavin is a complex drug including riboxin, riboflavin, nicotinamide, succinic acid. In recent years the use of cytoflavin has become widespread in the complex therapy of several diseases. In patients with acute infectious pulmonary destruction the inclusion of antihypoxant, antioxidant cytoflavine in the complex therapy contributed to a favorable course of the disease, rapid positive dynamics of the parameters of systemic inflammatory response and oxygen homeostasis, which contributed to the reduction of patient treatment period [60].

It was noted that during experimental brain ischemia in rats there is activation of glycolytic reactions, suppression of tricarboxylic acid cycle, due to insufficient brain oxygenation, there is an increase in lactate level. Cytoflavin had a detrimental effect on metabolic disorders, decreased lactate content, and its normalization was observed [59].

In chronic sensorineural hearing loss, citoflavin has been used as a remedy for regular courses of supportive (otoprotective) therapy. Cytoflavin in this case had a nonspecific central neurotropic effect and promoted the complexation of hearing function with peripheral deficits [61].

In acute renal ischemia, there was an increase in sympathetic activity and norepinephrine release from nerve endings. The use of maxonidine in renal ischemia influenced 1 -imidazole receptors and regulated sympathoinhibition [62].

The use of the complex preparation cytoflavin accelerated the normalization of myocardial energy supply [60]. In acute myocardial ischemia and bacterial endotoxin shock the antioxidant and antihypoxant activity of cytoflavin was revealed. The researchers explain the detected effects by the mutually potentiating effects of succinic acid, riboxin, riboflavin and nicotinamide [63,64].

Ketamine in cats had a sympathomimetic effect by increasing norepinephrine outflow induced by myocardial ischemia [65].

In rats that underwent myocardial infarction after coronary artery ligation, there was detected renal sympathetic nerve activity and contraction of the tibia trunk muscle. Intraarterial injection of kinin, which is a receptor B-antagonist, significantly decreased shin muscle contraction [66].

In acute myocardial ischemia in dogs caused by coronary artery ligation, impaired signal transmission during left sympathetic nerve stimulation was detected [67].

In acute experiments in pigs after coronary artery ligation, 75% of animals died for the first 30 min as a result of ventricular fibrillation of the heart, the other 25% lived 60 min - myocardial ischemia. In the animal, a dramatic decrease in norepinephrine content from $1.25\pm0.2\%$ to $0.67\pm0.10\%$ was observed in cardiac tissue during the first 15 minutes, whereas the initial norepinephrine concentration was 612 ± 72 to 402 ± 64 ng/g of crude tissue weight during the first 5 minutes. Researchers have concluded that in a model of myocardial ischemia, early ventricular fibrillation development is associated with norepinephrine release from sympathetic neurons after the onset of myocardial ischemia [68,69].

The development of ventricular arrhythmia and the decrease in norepinephrine in cardiac tissue 226 ± 77 versus 733 ± 82ng/g in myocardial ischemia in dogs has also been confirmed by other studies [70]. And there is also an opinion that in myocardial interstitial fluid and in the left circumflex coronary artery the concentration of catecholamines remains unchanged .

Coronary artery ligation caused loss of norepinephrine and adrenergic nerve endings in myocardial ischemia. The released catecholamines activated myocardial adrenoreceptors, which is associated with ischemic damage and ventricular arrhythmia [71]. In the outer shell of the epicardium in close proximity to the ventricular cavity during myocardial ischemia, floppy nerve terminals were detected. Around small arteries and arterioles a zone with stained fluorescence of nerve fibers was observed [72].

In rats with tibial nerve ligation, a disruption of the adrenergic innervation of small and medium arterioles was detected after 3 days, and after 7 days the fibers around the large arterioles were not visible. No fluorescent fibers were detectable even 84 days after surgery. It was concluded that tibial nerve ligation leads to a complete irreversible loss of adrenergic regulation of limb blood flow [73].

It has been shown that chronic arterial circulatory disorders in the lower extremities result in the accumulation of lipid peroxidation products in tissues [74].

It has been noted that the lymphatic system plays an important role in the regulation of oxidative homeostasis in norm and in circulatory disorders. It has been revealed that lymph nodes in ischemia are able to regulate oxidative homeostasis and maintain drainage and detoxification function [11].

There are no studies in the available literature devoted to adrenergic innervation of lymph nodes and vessels in chronic limb ischemia and its correction.

The popliteal lymph node in intact rats has an average diameter of 0.3-0.5 cm. Next to a large lymph node, there are several (2 to 3) small nodes. The number of arteries penetrating the lymph nodes is proportional to the size of the node.

In the tissue of the popliteal lymph node it was found that the greatest number of adrenergic nerve fibers is located near the follicular area of the lymph node, which form an adrenergic nerve ring innervating the follicles. The follicular innervation is formed mainly by individual single adrenergic fibers. In addition to the follicular adrenergic ring, there is an adrenergic plexus formed in the direction of the blood vessels feeding the lymph node tissue. Adrenergic nerve fibers belonging to blood vessels have more varicose thickenings and their fluorescence is higher, indicating a greater concentration of noradrenaline in these nerve fibers (Fig.17).

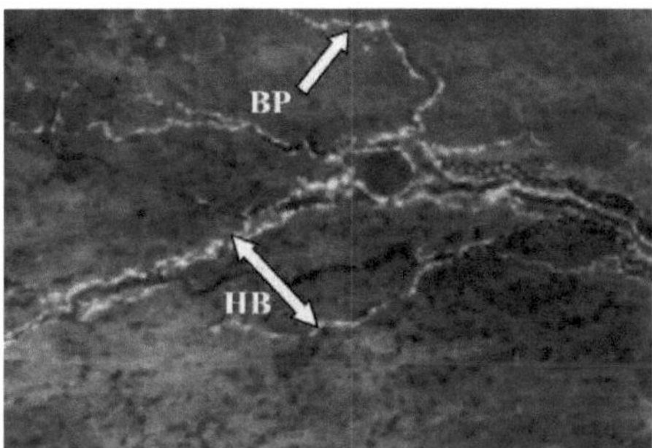

Figure 17 Adrenergic innervation of the capsule of the popliteal lymph node in the intact rat (the arrow indicates fluorescent nerve fibers - NV and varicose extensions -VR).

Ob.30. Ok, rk 6.3x

Thus, the tissue of the lymph node receives innervation to a greater extent through the adrenergic innervation of the blood vessels of the lymph node. This type of adrenergic innervation of the tissue is considered indirect. Consequently, the indirect type of vasomotor innervation predominates in the lymph node tissue.

Using cytoflavin in rats after 2 weeks of chronic posterior ischemia in the tissue of the popliteal lymph node a slight blood filling of the blood vasculature feeding the lymph node tissue is observed. Adrenergic nerve fibers (NV) accompanying vasa vasorum (VV) (Fig.18) and varicose dilation (VD) nerve fibers are diffuse. Independent thin terminals (TT) are preserved; however, the fluorescence of nerve fibers is weaker than in intact animals. The terminal varicosities become irregular in character and have a weak glow. Catecholamines released in the process of diffusion may apparently participate in activation of adrenoreceptors of the lymph node, which affects its functional state. The release of catecholamines we observed during the detected diffusion can activate the adrenoreceptors of the lymph node.

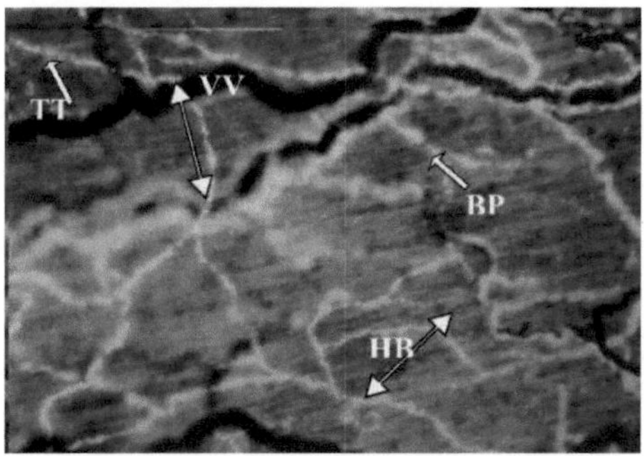

Figure 18. State of the adrenergic nerve plexus of the rat patellar lymph node after 14-day correction of hind limb ischemia. Release of catecholamines from adrenergic nerve endings, blood filling of microvessels of lymph node tissue is observed.

Ob.30. Ok, rk 6.3x.

In rats after 30-day correction of hind limb ischemia, near follicular nerve fibers were preserved in the lymph node tissue; however, compared to intact animals, even under correction conditions, they had weak fluorescence. Weakly fluorescent intermittent nerve fibers were observed in places. The bundles of nerve fibers in the adrenergic nerve network accompanying microcirculatory vessels had brighter fluorescence than independent nerve fibers. In the independent nerve fibers, in the preterminal part, varicose thickenings have a regular character compared to varicose thickenings in the terminal part. Terminal varices acquire a single character. Compared with the fluorescence of nerve fibers and varicose thickenings in the lymph node tissue in intact rats, the fluorescence of nerve structures in the lymph node tissue in rats with one-month correction of hind limb ischemia decreases by 30-40% (Fig.19)

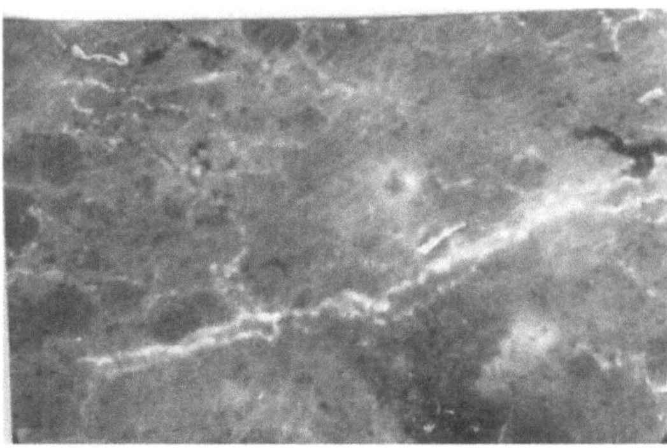

Figure. 19. State of the adrenergic nerve plexus of the rat popliteal lymph node
after 30-day correction of hind limb ischemia. There is a release of thecholamines
from the adrenergic nerve endings, blood filling of the microcirculatory
vessels of the lymph node tissue.

Ob.30. Ok, rk 6.3x.

In rats after 3 months of correction of hind limb ischemia, follicular single nerve fibers are preserved in the tissue of the hamstring lymph node. The circumvascular adrenergic nerve network is preserved to the extent that the microvessels in the lymph node tissue are preserved (Fig.20).

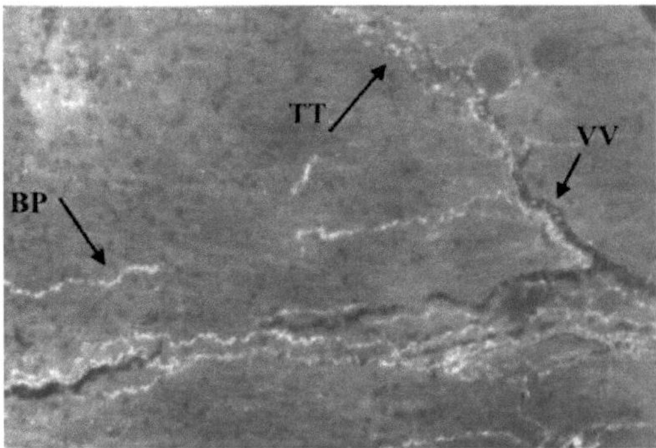

Notes: TT - subtle terminals ; VV- vasa vasorum; NV- nerve fibers; VP- varicose extensions.

Figure 20 Adrenergic nerve fibers in the
rat popliteal lymph node capsule
after 3-month correction of hind limb ischemia.

Ob.30. Ok, rk 6.3x.

Thus, intraperitoneal injection of cytoflavin, as seen from our data, prevents significant disruption of the structure of the adrenergic innervation apparatus of the lymph nodes and vessels. However, even with cytoflavin in the first 2 weeks there was a diffusion of catecholamines from the nerve structures. A decrease in fluorescence of the terminal nerve fibers was detected. Terminal varicose thickenings were irregular and had weak glow. After 30 and 90 days of correction of hind limb ischemia, the vascular adrenergic nerve network was preserved to the extent that the microvascular network in the lymph node and vessel tissue was preserved.

3.3.2. Peripheral blood flow and contractile activity of the cervical and popliteal
lymph nodes during cerebral and hind limb ischemia on
correction
with
semax or cytoflavine.

Along with cytoflavin after literature analysis we decided to take for correction of cerebral ischemia and cervical lymph nodes condition a drug developed at the Institute of Molecular Genetics of the Russian Academy of Sciences - "Semax". Semax is a synthetic analogue of ACTH, but without toxic and side effects and hormonal activity, it easily crosses the blood-brain barrier. In experiments on animals it was established that it possesses nootropic properties, increases adaptive abilities of a brain, has antioxidant, antihypoxic and neurotrophic effects. [75]
Semax has a strong complex neuroprotective effect, the main components of which are immunomodulation, inhibition of glial inflammatory reactions, improvement of brain trophic supply, inhibition of nitric oxide synthesis and oxidative stress reactions. Neuropeptide-induced chains of metabolic transformations reinforce and support each other, leading to the inhibition of most important mechanisms of delayed cell death.[76]
After occlusion of the carotid arteries or femoral artery, lymph flow from the jugular vessel and iliac lymphatic trunk of dogs decreased by 25 and 30%, respectively. When carotid arteries were occluded, blood and lymphatic clotting time decreased, while their viscosity, hematocrit, hemoglobin level, number of erythrocytes and platelets in blood increased, indicating increased thrombogenic processes. In lymph the number of lymphocytes increased, probably due to their release from lymph nodes. At brain reperfusion (60 min) the pH of blood and lymph decreased by 28 and 25%, respectively, i.e., toward acidosis. Blood oxygen tension decreased from 137 mm Hg to 78 mm Hg, indicating the onset of oxygen starvation in the tissues. The shifts of blood and lymph parameters in hind limb ischemia in rats were similar to those observed in cerebral ischemia. However, hypoxia and acidosis in hind limb ischemia-reperfusion were less pronounced. The contractile activity of cervical and popliteal lymph nodes in rats under tissue ischemia decreased, which was expressed in the reduction of amplitude and frequency of spontaneous contractions. The indices of hemodynamics in tibia arteries in intact animals were as follows: Vmax was 21.7±2.3 cm/s, V min 13.2±1.2 cm/s PI 0.63±0.02, RI 0.4±0.03. Comparative assessment of quantitative Doppler parameters of peripheral arterial blood flow in rats with chronic hind limb ischemia and control group. The results of Doppler studies in lower limb ischemia of different duration showed a tendency to decrease femoral artery blood flow. After 14 days the blood flow decreased to 7.38±1.2 cm/s. (Table 10)

After 1 month of ischemia, the pulsatility index increased to 1.63±0.01 and two months of ischemia to 1.05±0.02. Increased pulsatility index, characterizes arterial lesion in the lower extremities of the rat. According to the literature data, it is shown that with increasing degree of ischemia there is a decrease of pulsation index

Peripheral blood flow indices in rats of all groups at different duration of ischemia and correction with cytoflavine

Table 10.

Pok azat Spruce	inta ctn	14 days after ischemia	1 month after ischemia	2 of the month	correction Cytoflavin 3 days	14 days
V max	21.7 ±0,3	7.38±0,1	11,7±0,5	12,0±0,5	13,1±0 ,5	13,5±0 ,51
V min	13±1 ,5	4,78±0,2	7,81±0,2	10,2±0,4	12,5±0 ,4	10,2±0 ,3
Ri	0,4± 0,56	0,35±0,0 2	0,33±0,05	0,2±0,2	0,03±0 ,2	0,02±0 ,05
Pi	0,63 ±0,3	2,05±0,0 23	1,64±0,2	1±0,05	0,03±0 ,2	0,01±0 ,2
ISD	1,67 ±0,0 5	1,55±0,0 8	1,4±0,03	1,45±0,0 5	1,03±0 ,05	1,02±0 ,02
H R	180±2,3	99,7±0,2	100±0,2	105±0,5	108±0, 05	115±0, 2

From the presented data we can see that during ischemia-reperfusion of the brain and hind limb of animals the organ lymph flow decreases, associated with a decrease in lymphatic processes, as the increase in the proportion of form elements in the blood, as a compensatory response of the blood system in response to oxygen deficiency, needs an inflow of additional fluid into the bloodstream.

Administration of cytoflavin after limb ischemia 3-14 days by

showed an improvement in blood supply by 40-50% and an increase in lymph flow. (Fig.2,3)

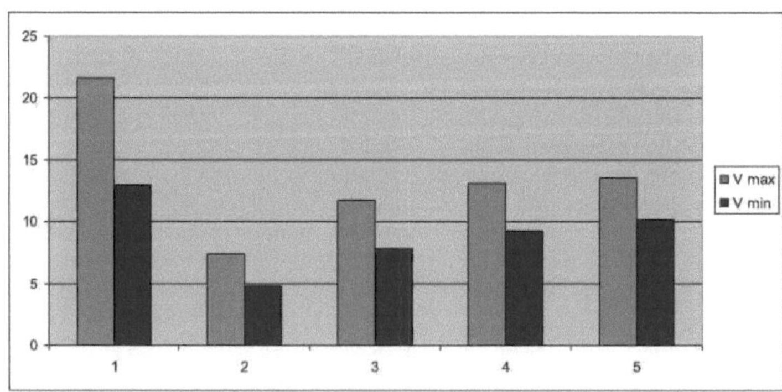

Рис 2.
Скорость
blood flow before and after 3,14 days of ischemia and their correction
Notations: on the abscissa axis: 1-control,2-after ischemia 3 days,-3-after 14 days,4-after correction

3 days, 5-after 14 days.

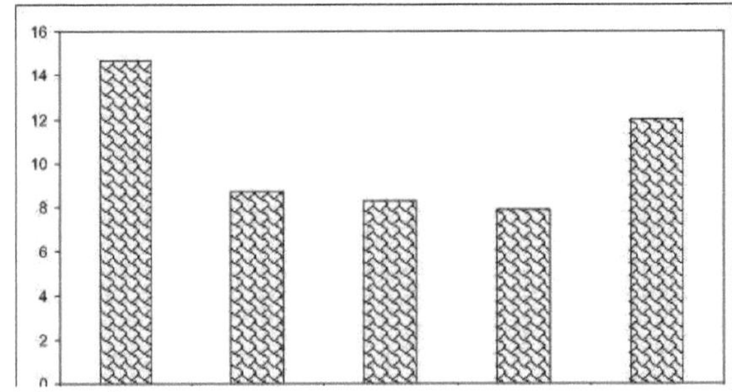

Fig.3.   Notations: by
ordinate axis - lymph flow in µl/min/100g body weight, abc axis - stages of experiments: 1 -
baseline, 2 - ischemia for 3 days, 3 - ischemia for 14 days, 4 - 60 days, 5 - application of cytoflavin
(after 60 days)

This figure shows that during ischemia of different duration the microcirculation of the
ischemized area worsens, the pulsation index decreases, and microvessels become structurally
deformed as a result of the disorders of their tone regulation mechanism. Hind limb ischemia
resulted in decreased blood supply of limb tissues, oxygen starvation and in parallel in decreased
lymph flow due to decreased lymph formation process, level of fluid resorption from interstitial
spaces, tissue pressure and capillary filtration.
As stated above, during organ ischemia - reperfusion the number of lymphocytes in lymph increases
and the latter, entering the blood, complement the form elements of the blood. On the other hand,
with a decrease in lymph flow and lymph formation associated with a decrease in resorption, tissue
fluid enters the bloodstream, as erythrocytes ejection from blood depot into the bloodstream during
circulatory hypoxia and hematocrit increase due to it, needs an inflow of additional fluid into the
bloodstream. Consequently, in circulatory hypoxia caused by ischemia-reperfusion of the brain and
hind limb, lymphodynamics is depressed and rheological properties of blood and lymph are
impaired. The lymphatic system, interacting with the blood system, participates in the regulation of
compensatory-adaptive processes aimed at the elimination of oxygen starvation in the tissues
caused by circulatory hypoxia in the area of the brain and hind limb.

The contractile activity of rat cervical and popliteal lymph nodes decreased during tissue
ischemia, which was expressed in a decrease in the amplitude and frequency of spontaneous
contractions.
In intact rats there was clearly pronounced spontaneous contractile activity of isolated
cervical lymph nodes in the form of phasic rhythmic contractions with a frequency of 3.0±0.4
socr./min and amplitude of 8.0±0.3 mg and hamstring lymph nodes was expressed in frequency -
3.2±0.2 socr./min, amplitude - 5.1±0.4 mg.
In them, contractile reactions of the cervical lymph nodes were observed in response to the
action of vasoactive substances. Adrenaline at a concentration of 10-8M-10-3 M when applied to
the cervical lymph nodes caused a significant increase in their rhythm by 162±12% and an increase
in the amplitude of contractions by 129±17% from the initial background. Acetylcholine (10-8M-
10-3M) caused a 142±13% increase in node contraction rate and a 130±12% increase in amplitude
from baseline. Histamine (10-8M-10-3 M) caused similar reactions when acting on cervical lymph
nodes. The threshold dose of vasoactive substances to induce node contractile reactions in intact
rats was 10-8 M.
Ischemia-reperfusion of the brain (after 1 hour from the start of reperfusion) in rats caused
suppression of spontaneous contractile activity of the nodes and only 55% of experiments showed
spontaneous rhythmic or tonic activity. The frequency of node contractions was 2.0±0.1 cr/min, and

the amplitude was 3±0.2 mg (Fig. 21). Induced contractile reactions of cervical lymph nodes in response to vasoactive agents at a concentration of 10^{-4}-10^{-6} M were observed in 50% of the experiments.

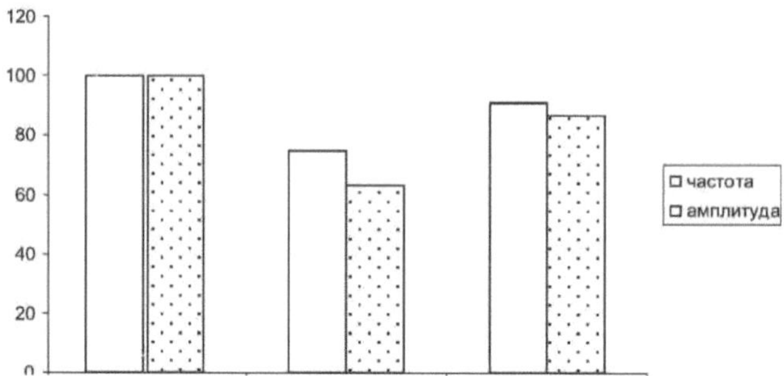

Notations: on the ordinate axis - shifts of node reduction in %, on the abscissa axis: 1, 2, 3 - groups of animals.

Fig.21. Frequency and amplitude of spontaneous contractions of cervical lymph nodes in control (group 1, taken as 100%), after brain ischemia (group 2) and against the background of semax correction (group 3).

After carotid artery occlusion, node contractile reactions under the action of adrenaline were detected in 55% of the experiments with a decrease in both the amplitude and frequency of node contractions. The amplitude of the contractions decreased by 20% and the frequency by 46% compared to the reactions observed in intact animals (Fig.22)

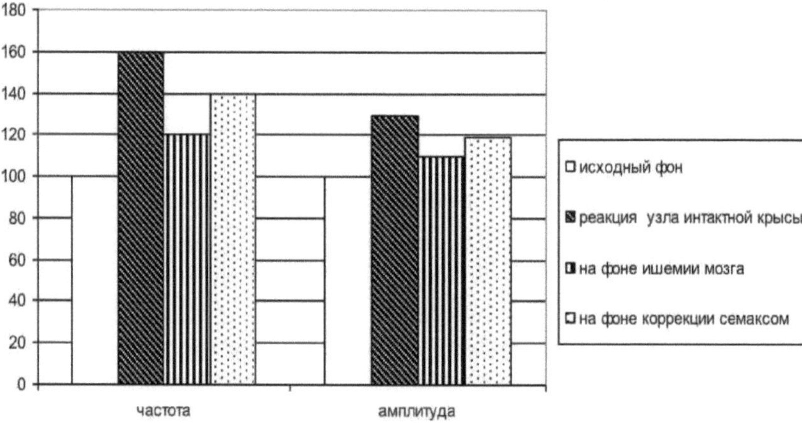

Notations: the first column is the initial background taken as 100%, the second column is the node reaction in intact rats, the third column is against ischemia-reperfusion of brain and the fourth column is against semax correction. Notations: the ordinate axis is the magnitude of shifts of contractions in %, the initial one is taken as 100%; the abscissa axis is the frequency and amplitude of contractions.

Fig.22 The contractile response of the cervical lymph nodes to the action of adrenaline (1x10-5 M).

Staining reactions of cervical lymph nodes in response to acetylcholine action were detected only in 35% of experiments and their magnitude was reduced. Both frequency (by 40%) and amplitude of contractions (by 43%) of smooth muscle cells of lymph nodes decreased, especially during brain reperfusion. Histamine, when applied to the node, caused contractile reactions of the nodes in 38% of the experiments: the amplitude and frequency of contractions decreased (Fig.23). Threshold doses of vasoactive agents to induce node contractions after brain ischemia-reperfusion increased by two orders of magnitude, up to 10-6M compared to the control value of 10-8M.

In 10 experiments, animals subjected to cerebral ischemia-reperfusion and treated with Semax according to the above scheme suppressed the spontaneous contractile activity of the cervical lymph nodes to a lesser extent than those rats who did not receive the above drug. However, the magnitude of induced node contractions in response to vasoactive agents after Semax correction was 10-12% lower than in intact rats (Figs. 23, 24).

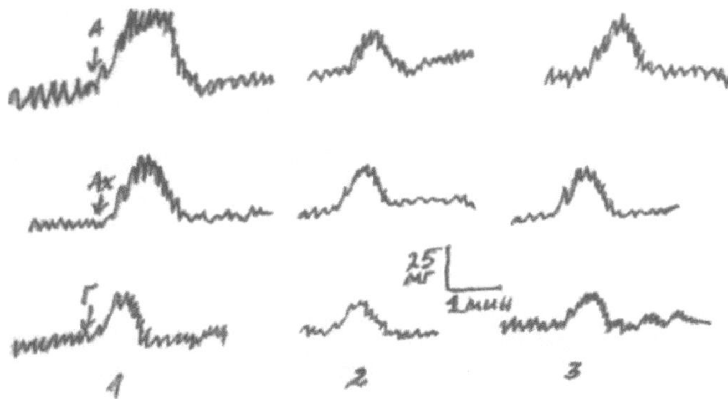

Figure 23. Contractile reactions of cervical lymph nodes before, after cerebral ischemia-reperfusion and semax correction. Notes: 1 - node contractile reaction in intact rat, 2 - node reaction against the background of cerebral ischemia-reperfusion, 3 - after giving semax., A - adrenaline, Ah-acetylcholine, H-histamine (1x10-6, 1x10-4M). ∎

The study of spontaneous contractile activity of the hamstring lymph nodes in 14 days from the beginning of hind limb ischemia revealed a decrease in the frequency and amplitude of contractions in 62-65% of experiments. The frequency of contractions of the hamstring lymph nodes was 1.3±0.1 socrine/minute, the amplitude decreased significantly, almost 1.5 times from the initial values and was - 3.5±0.1 mg. The contractile activity of the papillary lymph nodes under the action of the vasoactive agents adrenaline, acetylcholine and histamine (1 x10-4M-1*10-3M) was reduced by 40-50% compared to control (Fig.24).

According to the results of studies on rats with femoral artery occlusion in the groin area of the right hind limb, activation of spontaneous contractile activity of the popliteal lymph nodes was observed after 30 days. The frequency and amplitude of node contractions in the relative resting state were 20-25% higher than in intact rats. However, in response to the vasoactive agents adrenaline, acetylcholine, and histamine (1*10-6M-1x10-3M), the evoked node contractile activity was either reduced by 5% to 10% in 60% of the experiments, or the pattern of node contractions was similar to intact rats. In rats 3 months after the onset of posterior femoral artery occlusion, only 20% of experiments showed a slight decrease in the evoked contractile activity of the hamstring lymph nodes. The expression of contractile reactions decreased and their latency period increased.

During experimental ischemia, cytoflavin was administered after 14 and 30 days.

After 14 days of cytoflavin administration, the frequency of contractile activity of the hamstring lymph nodes was 2.0±0.1 socr/min and the amplitude 4.0±0.1 mg. The contractile activity of the lymph nodes on the action of adrenaline increased: amplitude by 20% and frequency by 15% (p<0.01), but were lower than in the control group. Acetylcholine and histamine in 70-80% of the experiments caused contractile reactions of the nodes with a decrease in their frequency and amplitude. When histamine acted on the nodes, their relaxation was observed in a number of cases. The stimulus threshold for vasoactive substances in ischemia in rats increased up to 1x10-4M. More than 1 month after cytoflavin correction, the evoked contractile activity of the hamstring lymph nodes on adrenaline action decreased in amplitude (by 40%) and frequency (by 30%), the threshold doses increased by 1x10-7M compared to the control value of 10-8M (Fig.24).

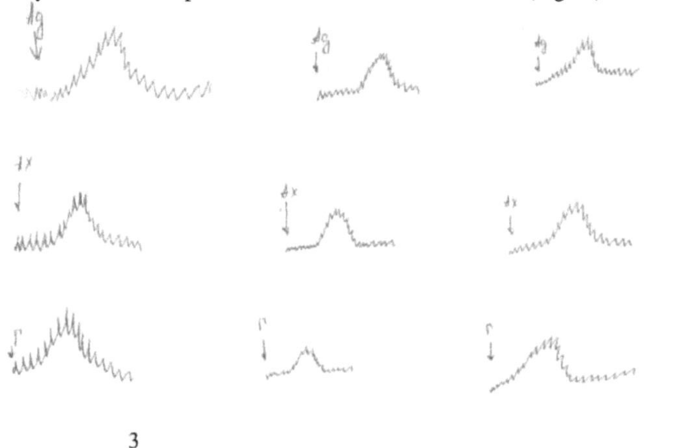

Notes: 1 - node contractile response in intact rat, 2 - node response against ischemia in 14 days, 3 - after cytoflavine application. Ad-adrenaline, Ach-acetylcholine, H-histamine (1x10-6' 1x10-4M).

Figure 24. Spontaneous and induced contractile response of the saphenous lymph node.

The oxygen tension in the blood decreased insignificantly, by 20-25% of the norm after 30 days. (Fig. 25). When cytoflavin was used, oxygen tension in the blood decreased only by 10 - 15% of the norm

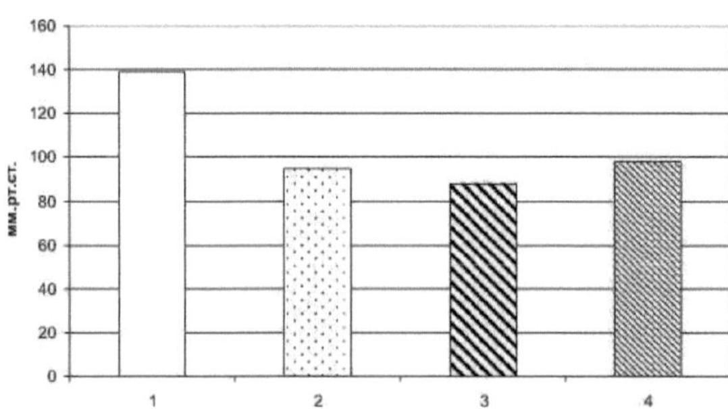

Designations: 1 - normal, 2 - after 30 min, 3 - 30 min reperfusion, 4 - 14 days after occlusion

Fig.25 Blood O2 tension in rats with femoral artery occlusion in the groin area.

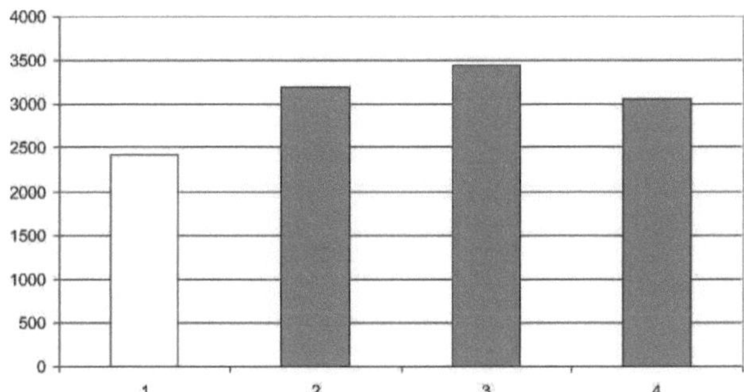

Designations:! - normal, 2 - 14 days after ischemia, 2 - after 30 days, 3 - 60 days after ischemia.
Fig.26 Changes in the number of lymphocytes in the lymph in lower limb ischemia.

An increase of 33% in the number of lymphocytes was detected in the lymph. This fact can be assessed as a compensatory reaction of lymph nodes that release lymphocytes into the lymph and deliver them into the bloodstream in response to tissue hypoxia (Fig.26).

The normal number of leukocytes in dry blood smears was 4 thousand, of which neutrophils were stabular - 6%, segmented - 67%, monocytes - 9%, lymphocytes - 18%. (Figure 27). The number of lymphocytes, monocytes, neutrophils - bacilliform and segmented neutrophils prevailed in the normal dry lymph smears. Lymphocytes accounted for 96% of the total number of cells in the lymph. In 14 days after lymphatic obstruction of the hind limb, the number of neutrophils (stab neutrophils-7%, segmented neutrophils-76%) increased (Figure 28).In addition, monocytes were found to be 7% and promotonocytes (5%), lymphocytes were 5%. In the lymph, there was a 2% decrease in the number of lymphocytes and a slight increase in the number of monocytes and neutrophils.

CONCLUSION

The following conclusions are drawn based on the data obtained:
1) Short-term ischemia-reperfusion of the brain and hind limb in rats and dogs is accompanied by disturbance of rheological properties of organ lymph, which was expressed in acceleration of lymph clotting time and increase of its viscosity, increase of thrombogenic processes in blood, increase of lymphocyte number in organ lymph, phenomena of acidosis as well as changes in biochemical parameters of lymph.
2) In case of cerebral and posterior limb ischemia already in the first minutes, according to rheography and dopplerography, there is a sharp decrease in blood supply of tissues, organ lymph transport, spontaneous and induced contractile activity of cervical and popliteal lymph nodes.
3) In rats with chronic ischemia of the hind limb, disruption of the adrenergic nerve network and structure of the nerve bundle in the tissue under the knee lymph node and vessels, both blood and lymph vessels, was noted.
4) Prolonged hind limb ischemia in rats was accompanied by decreased lymph flow, changes in biochemical and rheological parameters of lymph, prolonged decrease in blood supply to the hind limb, spontaneous and evoked contractile activity of the patellar lymph node.
5) Methods of prevention and correction of cerebral and hind limb ischemic injuries by using Semax for correction of cerebral ischemia and Citoflavin for hind limb oxygen starvation were developed.

LIST OF REFERENCES

1. Borodin Y.I., Grigoryev V.N. Lymph node in circulatory disorders.// Novosibirsk, -1986. -268.c.
2. Bulekbaeva L.E. Demchenko G.A. Transport function of lymphatic system in circulatory disorders // Proceedings of the MES of the National Academy of Sciences of Kazakhstan. Ser. Biol. and Med. 2001 - .№4, -C.30-34.
3. Astashova T.A., Kazakova E.S. The role of lymphatic system in regulation of oxidative homeostasis in norm with circulatory disorders and correction // Problems of lymphology and interstitial mass transfer: Mater.
4. Demchenko G.A. Lymphodynamics and vascular permeability in experimental hypertension // Medits. Journal. Astana, 1999.-#1, -P.13-17.
5   Bulekbaeva L.E., Demchenko G.A., Abdreshov S.N. Functional state of lymphatic system when simulating physiological effects of weightlessness // In the book: "State and prospects of scientific and innovation activity in the space sphere of RK". Almaty, 2005. -C.502-506.
6 . Petrishchev N.N., Vlasov T.D. Functional state of endothelium during ischemia-reperfusion // Sechenov Physiol. - 2000.- VOL.80, №2.-P. 148-161.
7   Vlasov T.D., Vivulanets E.V., Mendukshev I.V., Petrishchev N.N., Tvereva E.K. Functional activity of platelets at ischemia/reperfusion of rat brain // Physiol. zhurn. - C.422-426
8. Kirheby Ole J., Kutszsche Stefan, Risoe Culie Rise Ingum R // Cerebral nitric oxide concentration and microcirculation during in pigs // J. Clin Neuroschi.- 2000.- V. 7, № 6, P. 531-538
9. Bruce Jason I.E., Austin Cldre. Mechanisms of hypoxic vasodilatation in rat mesesenteric arteries: Role of intracellular calcium // J. Physiol. Proc. - 2000.- 523, P. 118 - 119.
10. Neumor Robert W. Molecular mechanisms of ischemic neuronal injury // Am . Emergency Med.- 2000.- 36, № 5, P. 483-506.
11.. Aubakirova H. J., Beketayev A. M. Influence of circulatory type of oxygen deprivation on lymph flow . Proc. Inst. of Physiol. of the Academy of Sciences of the Kazakh SSR   ,
Alma-Ata. vol. 8. 6-16. 1968
12 Zhumadina ShG, Bulekbaeva LE Comparative physiological study of lymph and hemodynamics in circulatory cerebral hypoxia in lower vertebrates. Bulletin. KazSU. № 3. 57-62. .2001.
Zhumadina Sh.M., Bulekbaeva L.E., Development of mechanisms of regulation of hemo- and lymphodynamics in phylogenesis of positional animals, Almaty, 2007. 172 c.
14.. Bulekbaeva L.E., Demchenko G.A., Vovk E.I. Lymphodynamics at ischemia reperfusion of canine small intestine // Ross. I.M. Sechenov. -2005,- №9.-C. 1066-1069.
15. Novikov I.P. Study of anatomical and functional changes in the lymphatic system in peripheral circulation disorders by means of lymphography.-Expert. Chir., 1963, ¹2, pp.3-6.
16. Prokofiev V.F. Lymph nodes in arterial ischemia. Novosibirsk, 1976, p.143-144.
17. Zorina S.V. Intestinal lymphatic channel in case of blood flow disturbance in experiment.-Arch. anat., histol. and embryol. 1972, No.9, pp.50-55
18. Hochachka P.W. Evolution of hypoxia tolerance: diving pinniped model and human hypobaric hypoxia model // XXXIII International congress of physiological sciences. - 1997, L.072.07.
19. Khitrov N.K., Paukov V.S. Adaptation of the Heart to Hypoxia. - Moscow: Nauka. 1991- 240 c.
20. Agadzhanyan N.P., Mirrakhimov M.M. Mountains and body resistance, Moscow, 1970, 184 p.
21. Boudko A.A. Patterns of breaching during adaptation under circulation hypoxia Conditions // XXXIII International Congress of physiological sciences. - St.
22. Kolchinskaya A.Z. On the role of oxygen and carbon dioxide in the regulation of interaction between respiration, blood circulation and erythrones // Oxygen regime of an organism and mechanism of its provision: Theses of All-Russian Conf. Barnaul, 1983.- P.20-21.

23.   Tkachenko B.I. Physiology of blood circulation. Physiology of Vascular System. Guidelines on Physiology. L.: Publishing house "Nauka" 1986.- 652 p.

24.   Zapata P., Larrain C. The role of dopamine in arterial chemo receptors: a controversial issue // XXXIII International Congress of Physiological sciences. - St. Petersburg, 1997, L. P. 030.23.

25.   Villanueva S., Mosgueira M., Inturriaga R. Dial effect of nitric oxide on carotid body chemo receptors // Abstr. Sci. Med. Phesiol. Proc. - 2000.-V. 523, P.109.

2 6.Oxman T., Arad M., Klein R., Avazov N., Limb ischemia preconditions the heart against reperfusion tachyarrhythmia. Amer. Y. Physiol. 1997, V.273. N 4, P.1707- 1712.

27.   Johnson Christopher D., Areher Joanne Morshall Janice. The role of sympathetic nerve activity in responses evoked in the rat tail artery by systemic hypoxia: Simultaneous recordings of neuronal activity and blood flow // Abstr Jt. Sci Mut. Physiol. Soc. With Brit. Pharmacology Soc. Southampton, Sept. - 1998, P. 8 - 11.

28.   Marshall, Janicem. Blood Flow regulation in systemic hypoxia // XXXIII International Congress of physiological sciences.- St. Petersburg,, 1997, L.056.03

29.Obreztchikova M., Taracova O., Koshelev V. Exploration of the pulse interval spectral powers in rats preliminary trained to hypobaric hypoxia // XXIII International Congress of physiological sciences.- St. Petersburg, 1997, P. 41 - . 51.

31.   Tabarov M.S. Adrenergic reactions of arterial and venous vessels of skeletal muscles of a cat at hypoxia and hypothermia // Physiol.zhurn. 2001,-87, ¹1,- P.43 -49. 2001 г.

32.   Agadzhanyan N.A., Efimov A.I., On the role of chemoreceptors in adaptation to hypoxia // Molecular aspects of adaptation to hypoxia. Collection of scientific works, 1979. 117.

33.   Shutov A.A., Chudinov A.A., Karaulova Y.V. State of hemostasis in ischemic stroke patients // Neuropathology and psychiatry. - 1991, 91, 7.- C.57-60.

34.   Varazashvili M.N., Mendelishvili G.I. Hematocrit in brain vessels in norm and its changes under ischemia // Theses of the reports of II All All conf. II All-Union Conference on Physiol, Pathophysiol, and Pharmacol, Tbilisi, 1988.

35.   Berezovsky V.A. Tension of oxygen in tissues of animals and man.-Kiev: Naukova Dumka. 1975.- 240 c.

36.   Higuchi Yuzo. Influence of arterial occlusion on hematocrit and plasma protein concentration of femoral venous blood in rabbit // Jap. J. Physiol.- 1985.- V.35, № 3, P.503 - 511.

37.   Lal Horbans, Williams K. Ivor, Woodword Brian. Chronic hypoxia differentially alters the responses of pulmonary arteries and veins to endothelia -1 and other agents // Eur. J.Pharmacol.- 1999.- 371, № 1, P.11-21.

3 8.Zubairov D.M., Kurochkin V.I. Coagulation and protein composition of lymph during acute blood loss // Bulletin of Expert Biol. and Med.-1962.-T.57, ¹ 3.- P.212- 217.

39.   Baizhanova N.S. Changes of hemo- and lymphodynamics in the early period of circulatory hypoxia: dissertation of doctoral thesis - Almaty, 1988 - 24 p.

40.   Kazakov O.V., Astashev V.V. The study of osmolar concentration of serum lymph and muscle tissue in the dynamics of ischemia with limb reperfusion // Proc. Int. symposium on clinical and expert. Lymphology. - Novosibirsk, 1998.- P.142 - 143.

41.   Satpaeva H.K., Krichevskaya I.P., Nildibaeva Z.B. et al., Functional changes in blood and lymph circulation systems under hypoxia // XIV Congress of All-Paul Pavlov Society of physiology: Collection of scientific works, L. , 1983.-T. 2.- C. 288 -289.

42.   Putalova I.N. Structure of heart lymphatic region - an indicator of morphofunctional state of cardiac muscle // Morphology. - 1998.- 113, № 3.- C. 9 8.

43.   Willson W., Shults R., Cooper J. The isolation of cholinergic synaptic vesicles from bovine superior cervical ganglion and estimation of their acetylcholine content // J. Neirochem.. - 1973.- V.20, P.659- 667.

4 4.Orlov R.S., Lobov G.I., Borisova R.P. Erofeev N.P. en al . Lymph transport theory and practice. // XXXIII International congress of physiological sciences.- St.Petersburg. .- 1997. - L035.08

45.   Gladysheva N.L. Effect of combined action of hypoxia and hypothermia on thoracic duct

contractile activity and portal veins // Mater. II Conf. of Young Scientists on Research Ser. su. Sist. - L., 1981.- P.29-31.

46. Caudill Timothy K, Resta Thomac C., Kanady Nancy L. et al. Role of enadhelial carbon monoxide -CO in attenuated vasoreactivity following chronic hypoxia // Amer J.Physiol.- 1998.- V. 285, № 4, Pt.2 , P.1025 - 1030.

47. Lobov G.I., Kubyshkina N.A. Effect of acidosis on the contractile function of bovine mesenteric lymphatic vessels // Bul. Exper. of Biol. and Med. - 2001.- 132, № 7.- C.16-19.

48. Herrera Gerald., Walker Benjimen R. Jnvolvemennt of L -tape calcium channels in hypoxic relaxation of vascular smooth muscle // J. Vass. Res. ,1998, V. 35, № 4, P. 265 -273.

49. Bruce Jason I.E., Austin Cldre. Mechanisms of hypoxic vasodilatation in rat mesesenteric arteries: Role of intracellular calcium // J. Physiol. Proc.- 2000.- 523, P. 118 - 119.

50. The autoregulation of cerebral circulation in vertebrates // Proceedings of IX All-Soviet Council on Evolutionary Physiol. - L., 1986. 142-143.

51. Brown V.W., Evans B.K. The bar receptor heart reflex of the toad (Buffo marinus) // Proc. Autral Physiol and Pharmacol Soc.- 1988.-V. 19, P. 38-48.

Krivchenko (47)

52. Garland T.J. Phylogenetic approaches in evolutionary physiology/ XXXIII International congress of physiological sciences-St.Petersburg , 1997. - L.072.04.

53. Kupriyanov S.I., Semenova M.K. Physiological characteristics of the reflexogenic zone of vertebral arteries // Uspekhi Physiologicheskikh Nauk. - 1994, VOL.25, №3, P.70-78

54. Valeeva Z.T. On innervation of thoracic lymphatic duct of a dog and its reaction to some poisons // Jour. Pharmacology and toxicology. - 1948.- T. 9, № 5.- C.49-50.

55. Smirnov D.I. On reflex from superior vena cava to lymphatic vessels // Bulletin of Exepr. - 1955.- T. 39, № 6.- C.19-21.

56. Kamyshnikov V.S., Kolb V.G. Clinical Biochemistry. - Moscow, - 2000,   - T. I-II, - 480 p.

57. Govyrin V.A., Leontieva G.R., Prozorovskaya M.P., Reidler R.M. Adrenergic nerves and venous catecholamines // I.M. Sechenov Physiol. zhurn. of USSR - 1981. - T.67, №1. - C.13-22.

58. Blattner R., Klassen H., Denert H. Experiments on isolated smooth muscle preparations. - Moscow. Mir. - 1983. - 206 c.

59 Bisenkova MN, Romantsov MG, Afanasyeva GA, Chesnokova NP Cytoflavin as a drug effective correction of metabolic disorders in hypoxia of various genesis // Adv. Natural Science. 2006,-№ 4- C.28

60 Chesnokova N.P., Romantsov M.G. Pathogenetic substantiation of cytoflavin use in ischemic myocardial damage // Fundamental'nye issledovanie. 2006, №4, C.20-25

61. Zhuravsky S.G., Romantsov M.G. The place of cytoflavin in hearing therapy for chronic sensorineural hearing loss // Modern Science-Intensive Technologies 2005. - № 9.- C.16-18

62. Bouillon VV, Khnychenko LK, Sapronov NA, Kovalenko AL, Alexeeva LE, Romantsov MG, Chesnokova NP, Bizenkova MN Metabolic effects of cytoflavin and piracetam in acute experimental brain ischemia and during its reperfusion // Advances of modern natural science. 2007.-№3, C.74-78

6 3 Belskih A.N., Fufaev E.E.. Application of cytoflavin antioxidant in combination with extracorporeal hemocorrection in patients with acute pulmonary suppurations // Bulletin of intensive care. 2007.№2.C.75-79.

64. Tsusui H. ,Sugira T. Hayasi K., Ohkita M., et al Moxidine prevents ischemia/reperfusion-induced renal injury in rats// Eur J Pharmacol.2009. 603(1 -3).P.73-78

65. Kitagawa H., Yamazaki T.,Akiyama T., Yahagi N. et al Modulatory effects of ketamine on catecholamine efflux in vivo cardiac sympathetic nerve endings in cats // Neurosci Lett. 2002. V 324, .№3 .P.232-236

66. Koba S., Xing J., Sinoway LI, Li J. Bradykinin receptor blockade reduced sympathetic nerve response to muscle contraction in rats with ischemic hearts failure // Am J Physiol Heart Circ Physiol. 2010. V.298, №5. P. 1438-1444

67. Martins JB., Kerber RE., Marcus ML., Laughlin DL., Levy DM. Inhibitions of adrenergic neurotransmission in is^aemic regions of the canine left ventricle// Cardiovasc. Res. 1980. V.14, №2. P.116-124.

68. McDonald F.M. , Knopf H., Hartono S. Polwin W ., et al. Acute myocardial ischaemia in the anaesthrised pig : local catecholamine release and its relations to ventricular fibrillation // Basic Res. in Cardiol. 1986.V.81,N.6. P. 636-645.

69. Roth D.M., White F.C., Mathieu-Costello O., Guth B.D., Heusch G., Bloor C.M., Longhust G.C. Effects of left circumflex Ameroid constrictor placement on adrenergic innervation of myocardium // Am J Physiol Heart Circ Physiol. 1987. V.253 № 6, P. 1425-1434.

70. Michael W. Dae, J William O'Connell, Elias H. Botvinick, Michael C.Chin. Acute and chronic effects of transient myocardial ischemia on sympathetic nerve activity , density, and norepinephrine content // J Med Cardiovasc. Res.1995.- V.30, Is. 2 .-P.270-280.

7 1Holmgren S. Abrahamsson T., Almgren O., Erikscon V.M. Effect of ischemia on the adrenergic neurons of rat heart fluorescence histochemical and biochemical studies // J Cardiovasc. Res.1981.-V.12, Is. 12 .-P.680-689

72. Abrahamsson T., Almgren O., Holmgren S. Effects of Ganglionic Blockade on Noradrenalin release and cell injure in the acutely ischemic rat myocardium // J Cardiovasc. Pharmacology.1982. V.4-P.584-591.

73. Koistinaho J, Wadhwani KC, Latker CH, Balbo A, Rapoport SI. Adrenergic innervation of blood vessels in rat tibial nerve during Wallerian degeneration. Acta Neuropathol. 1990; V.80 Is.6. P.604-10.

7 4.Ioskevich NN, Zinchuk VV Prooxidant-antioxidant balance in blood during the surgical treatment of obliterating arterial atherosclerosis in the lower extremities // Roczniki Akademii Medycznej w Bialymstoku -2004.-Vol. 49.-P.222-226

7 5Ashmarin I. P., Nezavibatko V. N., Myasoedov N. F., Kamensky A. A., Grivennikov I. A., Ponomareva-Stepnaya M. A., Andreeva L. A., Kaplan A. Y., Koshelev V. B., Ryasina T. V. Nootropic analogue of adrenocorticotropin 4-10- Semax (15-year experience of development and study) // Journal of VND, 1997, Vol. 47, vol. 3, p. 420 -430.

76. Gusev E.I., Skvortsova V.I. Ischemia of the brain. - Moscow: Medicine, 2001. - — 328 c.

77. Mayborodin I.V., Pavlyuk E.G., Shevela A.I. Variants of regional lymph node sclerosis in primary lymphedema // Proceedings of the International Symposium. Problems of lymphology and endoecology. - Novosibirsk,- 1 998.-P.257-258.

78. Meerson F.Z. Adaptation Medicine: Mechanisms and Protective Effects of Adaptation. MOSCOW: 1993, 253 PP.

79.Omarova A.S., Alibaeva B.N. Long-term consequences of stress effects of hypoxic factor on lymphatic circulation in rats. Mechanisms of functioning of visceral systems. YII All-Russian Conf. Conf. with International Participation, devoted to the 160th Anniversary since the birth of I.P. Pavlov, Saint Petersburg, Russia, 29 September-2 October 2009, P.324. 2009, P.324-325.

80. Zykov A.A. Phenomenon of lymphotropism from pharmacologist position // Bul. Novosibirsk, 2001, No 4 (October-December), Novosibirsk, p. 13-17.

81. Kiseleva T.L. Horse chestnut common . // Medical Aid. 1995. №2 C54-57

82. Bogachev V.Y. Modern pharmacotherapy of chronic venous insufficiency of the lower extremities // Pharmacological Newsletter. 2002, №12. C. 21-22.

83. Tikhonov AI, VA Soboleva, Pukhir NV Horse chestnut and its use in medicine.//Provizor, 1999, ¹ 22, P. 34 -37.

84. Frick RW. // Angiology 2000 .- V.51,-3:1.-P.97-205

85. Journal of the neurological sciences. V.58 Iss. 1 1983 P.37-44 biochemical changes during graded brain ischemia in gerbils PART 2 evaluation of cerebral blood flow and brain metabolites. W.Pashen, B.M. Djuricic, H.J. Bosma ,K.A. Hossmann

86. V.V. Pakhomova, A.V. Efremov. On the study of biometal concentrations in blood plasma, lymph and lymph nodes of rats in the dynamics of myocardial stress damage and against the background of selective enterodonorosorbent application// Bulletin of New Medical Technologies,

2009, Vol. XVI No. 2 P.32-34

87. Stureva G.M. Muslov S.A. Analysis of lymph properties in surgical pathologies // Fundamental Research. RAE #11 2007 P.23-25

88. N.I. Melnikova. Leukocyte-endothelial interaction in pial arterioles and venules during development of rat brain ischemia. Vol.94 No.1 -2008 P.45- 53

89. Lazarev A.M. Determination of blood leukocyte number in experimental intestinal ischemia against the background of blastocytosis. Collection of scientific works "Natural Science and Humanism. - Vol.5 vol.1.-2008, ed. prof. D.M.S. Ilyinskih N.N.

Lymphatic velocity in healthy and injured limbs. Sveshnikova K.A., Russeikin N.S. Scientific journal "Modern problems of science and education" №2 2008 RAE

90. V.I. Konenkov, V.F. Prokofiev, A.V. Shevchenko,E. V. Zonova. - Cellular, Vascular and Extracellular Components of Lymphatic System // Bulletin of Russian Academy of Medical Sciences,-#5 (133),-2008,-P.7-13

91. Astashova T.A., Antsyreva Yu.A., Kazakov O.V., Morozov S.V. Lymphatic system in the mechanism of oxidative homeostasis in modeling of circulatory disorders and their correction by low-energy laser radiation // Bulletin of Siberian Branch of RAMS, №1 (115).- 2005, P.74-78

92.ZaikoT   .M. Lymph nodes under conditions of general hyperthermia (morpho-functional study) :autoref. dissertation. c.m.s. -Novosibirsk, 1987.-17p.

93. A.V. . Efremov, K.K. Dmitrieva, A.V. Ignatova, S.V. Michurina, E.V. Ovsyanko Effect of controlled hyperthermia on morpho-functional state of visceral lymph nodes // Bulletin of new medical technologies.- 2008. - T 15. №3 - P.30-31.

94. Y.I. Borodin, T.A. Astashov, V.V. Astashov The role of lymphatic system in maintaining the mechanism of oxytocic homoeostasis in norm, in modeling of atherosclerosis and its enteric correction by sorbent // Bulletin of Siberian Branch of RAMS.-¹2.2006.- P.73-79.

95. I.V.Shipilova Ukrainian Journal of Extreme Medicine named after G.O.Mozhayev - T9, ¹3.- 2008 - P.98-101.

96. Harrison DC, Davis RP, Bond BC, et al. Caspase mRNA expression in a rat model of focal cerebral ischemia// Brain Res Mol Brain Res.- 2001.-89.-P.133-146.,

97. Bas O Songur A, Sahin O, et al. The protective effect of fish n-3 fatty acids on cerebral ischemia in rat hippocampus.// Neurochem Int.- 2007.- V.50.-P.548-554.

98. Oleg Y Grinberg1, Huagang Hou1, Stalina A Grinberg, Karen L Moodie, Eugene Demidenko, Bruce J Friedman, Mark J Post and Harold M Swartz. pO2 and regional blood flow in a rabbit model of limb *ischemia* //Physiol. Meas.- 2004.-V 25,- №3,- P. 659

99. Bizenkova MN, Romantsev MG, Chesnokova NP Metabolic effects of antioxidants in acute hypoxic hypoxia // Fundamental research .-#1, 2006.-p.17 -21.

100. Bouillon V.V. Khnychenko L.G. Sapronov N.A. During reperfusion of ischemized brain by the 3rd day its antiradical protection progressively decreases. Metabolic effects of piracetam in experimental cerebral ischemia and in the process of its reperfusion // Fundamental research. - №4.- 2007.- P.55

101. A.A. Toropova, S.V. Lemza, A.G. Mondoev, E.V. Petrov. - The influence of "nefrofit" on the dynamics of ATP content in the ischemic kidneys of white rats // Siberian Medical .-2009.- № 4.- P.115-116

102. Menshchikova E.B., Zenkov N.K., Lankin V.Z. et al.Oxidative stress (pathological conditions and diseases). - Novosibirsk. 2008. - — 283 с.

103. Bokeria L.A., Chicherin I.N. Nature and clinical significance of "new ischemic syndromes"- Moscow. A.M. Bakulev Research Center of Russian Academy of Medical Sciences, 2007. 302 p. Moscow.

104. Verma S, Fedak PWM, Eeisel R.D. Fundamentals of reperfusion injury for the clinical cardiologist// Circulation -2002. - — V. 105. - — P. 2332-2336.

105. Tkachenko B.I., Evlakhov V.I., Poyasov I.Z. Hemodynamic mechanisms of venous return

and pulmonary circulation parameters reduction at experimental myocardial ischemia.Bulletin of Experimental Biology and Medicine.-2009. Vol. 147 No.1 P.32-36.

106. N.V. Naryzhnaya, L.N. Maslov, Y.B. Lishmanov. Effect of adaptation to stress on the content of cyclic nucleotides in myocardial tissue during acute ischemia and reperfusion.// Bulletin of Experimental Biology and Medicine. No.5.-T.145,2008,-P.525-528 107 S.V. Aksenova, I.V. Atamakin, N.A. Ashirova, A.P. Vlasova. Aeroion therapy in the correction of homeostatic disorders of the intestine in acute ischemia // Bulletin of New Medical Technologies.- 2007.- T/XVL- №3.- P.131.

108. Aaslid R, Markwalder T.M. Nornes H. Non invasive transcranial Doppler ultrasound recording of flow velocity in basal cerebral arteries// J Neurosurg. 1982.- 57.-P. 769-74.

109. Halea A Corcoran; Brad E Smith; Parker Mathers; Dan Pisacreta; James C Hershey J. Laser Doppler imaging of reactive hyperemia exposes blood flow deficits in a rat model of experimental limb ischemia//Cardiovasc Pharmacol. V.53.-P.446-51

110. Vucaj-Cirilovic V, Nikolic O, Petrovic K, Govorcin M, Hadnadev D, Stojanovic S. Basic characteristics of duplex sonography in the assessment of lower limb arterial circulation //Med Pregl. 2006 V. 59.I.5-6.P.287-290.

111. A. Halea. Smith, E.. Brad , Mathers Parker , Pisacreta, Dan , Hershey, James C. Laser Doppler Imaging of Reactive Hyperemia Exposes Blood Flow Deficits in a Rat Model of Experimental Limb Ischemia Corcoran// Journal of Cardiovascular Pharmacology.- 2009 - V. 53 - I. 6 - pp 446-451.

112. Stroke: A Journal of Cerebral Circulation:V 29.- I.11.-November 1998.pp 2412-2420

113. Sarkisyan B.A., Velichko R.V. // Siberian Medical Journal. - Tomsk.-2008.- №1 P.1 P.42-43

  114. Yang S.,Chen B., Luo T.,Tong Z.,Zhang S. Construction and evaluation of rat hindlimb acute ischemia model temporal exposure of cryptic collagen epitopes within ischemic muscle during hindlimb reperfusion// American Journal of Pathology.- V. 167.- N.5, 2005.- P.13491359

115. Paul J. G, Nikita T, Xialou Li, Joseph G, Jhenrong Q, Michael S., H. Yee, Elizabeth G., and Peter Brooks Temporal Exposure of Cryptic Collagen Epitopes within Ischemic Muscle during Hindlimb Reperfusion. // Am J Pathol.- 1998.- V.-152 P.1667-1679

116. E.B. Artyushkova. Pashkov D.V., Myasnikov A.D. Possibilities of combined pharmacological and surgical correction of chronic limb ischemia in experiment. Authors' team 2008 P.1-4.

117. Yong L, Dingguo Z. Yuqing Z. Augmentation of neovascularization in murine hindlimb ischemia by combined therapy with simvastatin and bone marrow-derived mesenchymal stem cells transplantation//Journal of Biomedical Science.- *2010* V.17.P.75.

118. Ulrich, Peter T. Kroppenstedt, Stefan. Laser-Doppler Scanning of Local Cerebral Blood Flow and Reserve Capacity and Testing of Motor and Memory Functions in a Chronic 2-Vessel Occlusion Model in Rats

119. Paul J. G, Nikita T, Xialou Li, Joseph G, Jhenrong Q, Michael S, H. Yee, Elizabeth G, and Peter Brooks Mouse model of angiogenesis// Am J Pathol.-1998.-V.-152 P.1770-1775

120. D D Burks, B J Markey, T K Burkhard, Z N Balsara, M M Haluszka and D A Canning.Suspected testicular torsion and ischemia: evaluation with color Doppler sonography// Radiology. 1990.-V. 175.-P. 815-821

121. Federico B., Giuseppe S., Vincenzo A. Pioglitazone enhances collateral blood flow in ischemic hindlimb of diabetic mice through an Akt-dependent VEGF-mediated mechanism, regardless of PPARy stimulation// Cardiovascular Diabetology.- 2009.-V. 8.P.49

121 The effect of gradual or acute arterial occlusion on skeletal muscle blood flow, arteriogenesis, and inflammation in rat hindlimb ischemia//Journal of Vascular Surgery. - V. 41, I.2/- P. 312-320

122. Oleg Y Grinberg et al. pO2 and regional blood flow in a rabbit model of limb ischemia //Physiological Measurement. -2004 V. 25, N. 3_ 659

123. Bahri U. Ibrahim S. Mahmut K. Mustafa B.The Evaluation of Microanastomoses in Rats with

Color Doppler Ultrasonography Tr.// J. of Medical Sciences 29.- 1999.P. 249-252

124. Ultrasound and Vessels. Diagnostic practice. Disc under the editorship of O.Y. Atkov.

125. Campbell WB, Baird RN, Cole SE, Evans JM, Skidmore R, Woodcock JP. Physiological interpretation of Doppler shift waveforms: the femorodistal segment in combined disease. //Ultrasound Med Biol.- 1983-V.9,I.3. P.265-269.

yes I want morebooks!

Buy your books fast and straightforward online - at one of world's fastest growing online book stores! Environmentally sound due to Print-on-Demand technologies.

Buy your books online at
www.morebooks.shop

Kaufen Sie Ihre Bücher schnell und unkompliziert online – auf einer der am schnellsten wachsenden Buchhandelsplattformen weltweit! Dank Print-On-Demand umwelt- und ressourcenschonend produziert.

Bücher schneller online kaufen
www.morebooks.shop

KS OmniScriptum Publishing
Brivibas gatve 197
LV-1039 Riga, Latvia
Telefax: +371 686 204 55

info@omniscriptum.com
www.omniscriptum.com

Printed by Books on Demand GmbH, Norderstedt / Germany